Finding Gold in Colorado:
The Wandering Prospector

A sequel to *Finding Gold in Colorado:*
Prospector's Edition

...for Colorado gold panners and prospectors who want more dig sites!

Kevin A. Singel

Version 1.06
Copyright © 2023-2026 Kevin A. Singel
Photography by Laura A. Hoeppner
All rights reserved.
ISBN: 9798397578608 (paperback)
ISBN: 9798397815888 (hardcover)

DEDICATION

To my patient, creative, encouraging wife Laura.

To all the Colorado gold rush prospectors who came before us.

To those prospectors who bought my first book, showing me
there was a need for this book too.

And <u>especially</u> to those who are always eager to try a new spot!
"Find your gold!"

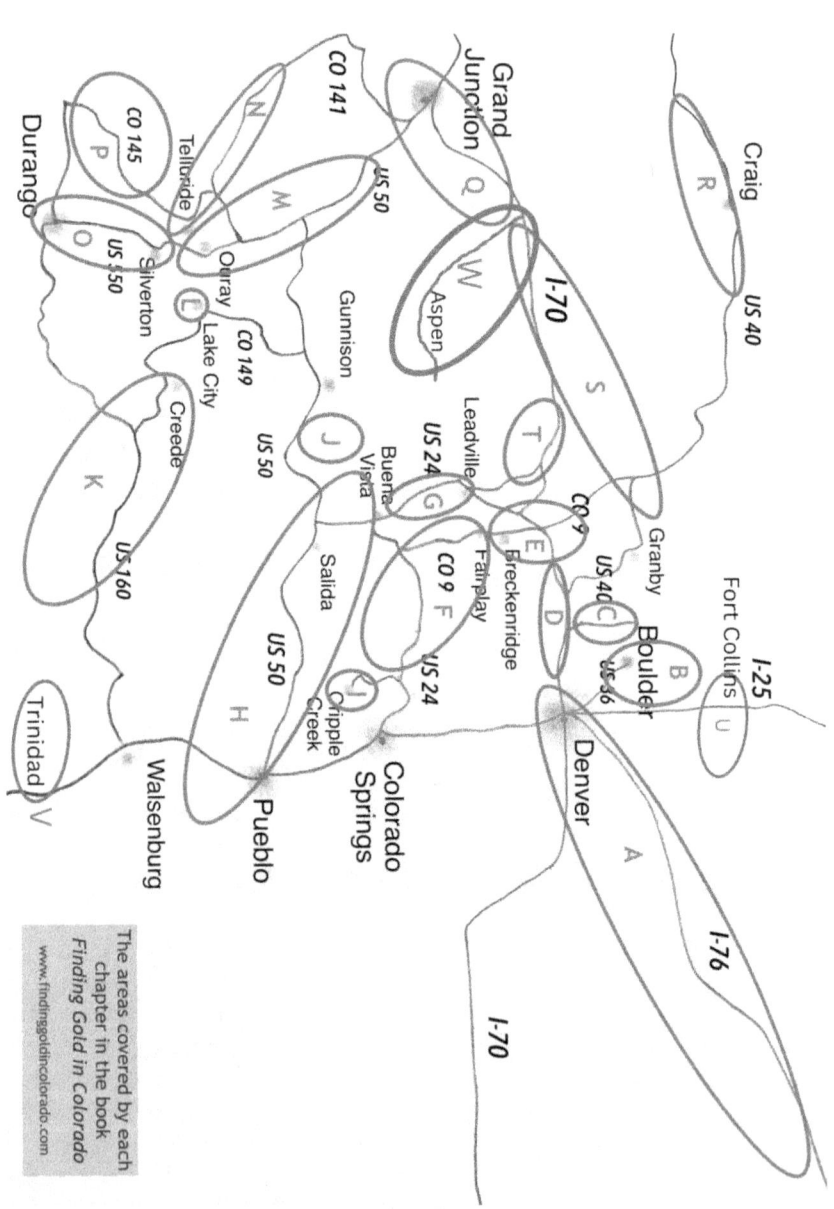

The areas covered by each chapter in the book *Finding Gold in Colorado*
www.findinggoldincolorado.com

TABLE OF CONTENTS

ACKNOWLEDGEMENTS

First, thanks to the members of the Finding Gold in Colorado group on Facebook. Their willingness to share information and their encouragement significantly expanded what was possible for me to include in this book. Second, thanks to all the folks who went into the field, on their own, or with me, to sample and explore potential prospecting sites. Your help and companionship in the field made this safer and more fun!

I'd also like to individually thank:

> My wife Laura Hoeppner for so many things – from being my travel buddy, to editing the final manuscript. She also contributed by taking the professional quality photos included here and in our other books, on the companion website, and published by both of us in the 'Finding Gold in Colorado' Facebook group.

> Jim Spinosa for his support in improving the companion website to this book, www.findinggoldincolorado.com, where you will find many pictures and more details on some of the sights and sites mentioned in the books.

> Dennis Henry for designing my logo, for creating his awesome Grizzly Goldtrap sluices, editing this book, and especially for the adventures we have shared with a gold pan in hand.

> Sue Hoeppner for hosting me for the writing retreat that allowed me to finish this book on time!

Thanks also to Mike Bobbit, Lucas Wentzel, Henry Willis, Alexandre Genovese, Doug Penninga and Dennis Henry for working with me these last few years as moderators of our Finding Gold in Colorado Facebook group (Hint: You should join the group, there are tens of thousands of helpful prospectors in there!)

And finally, a big thanks to Nick Hertz, Timm Kluender, Shane Schubarth, John Stanner, Brad Martinez, Brandon Lampley, Edward Vance, Matt Ingram, Chas Felthager, Jeffrey Nunn, David Boette, Jimmie Don Aycock, Timm Kluender, Logan Thacker, Marcus Bristow, Jeremy Toms, and Paul Schiffer for sharing their local knowledge and their help testing various sites to confirm safe access and gold of course! Paul's passion for the South Platte River in northeast Colorado and Jimmie Don Aycock's passion for the Conejos River are simply inspiring. Both of them eagerly take on the challenge of finding gold where other folks say it isn't! Oh, and my apologies to whoever I forgot to mention in this list. Drop me a note and I will add you in!

AUTHOR'S NOTE

The book in your hands is a follow-up expansion of *Finding Gold in Colorado: Prospector's Edition*. It is a sequel in the truest sense of the word, so I now refer to the books as Volume 1 and Volume 2. That said, this book can easily be used on its own, with no access to the original, although I definitely recommend beginners get the first book for or along with this one. Of course, the site numbering makes a LOT more sense if you also have the first book. In order to make it easy to talk about dig sites with other prospectors, I have decided to use the chapter structure from the first book in this one as well. So, for example, Chapter E in the prior book covered "Breckenridge and Summit County" so this one includes a Chapter E with the additional information I have discovered about that area. Similarly, the dig sites in this book start from the end of the numbering in each chapter of the prior book. So, to continue my example above, Chapter E in the first book included 4 dig sites so the dig sites described in this book's Chapter E start with E-05.

Many of the prospecting sites in this book are away from the famous gold mining towns and gold placer districts of Colorado. Some are in areas where there was minimal commercial mining. Others are in areas that were actively prospected and mined but never drew enough attention have a boom town of their own that lasted very long. As a result, there is less gold rush history and less to go see as a tourist.

Quite a few other sites included are in established placer districts which <u>were</u> covered in my prior book. These sites simply escaped coverage in that book for one reason or another – usually because I was unable to confirm legal access status before publication.

The gold at these sites may be sparse and barely worth chasing or surprisingly good. Either way, these are places where I think you will at least have a little fun with your gold pan if you are

in the area. Some of these sites surprised me with excellent gold and those sites may surprise you too when you try them!

If you'd like to get an impression of what these areas look like in full color, pick up a copy of our companion book *Finding Gold in Colorado: Inspiring Images*. It has the same chapter organization as this book (chapters A-T) and is loaded with full color photographs of Colorado gold country.

We also offer *Finding Gold in Colorado: Prospecting Log*. This well-formatted logbook gives you a structured way to record your prospecting outings. Doing this will help you keep track of where you have been, what you learned by prospecting there, and, of course, how much gold you collected!

Learn about all of our books and how to buy them at
www.findinggoldincolorado.com/fgicbooks/

INTRODUCTION

This is a guidebook to casual gold prospecting sites I have identified since the publication of my first book *Finding Gold in Colorado: Prospector's Edition*. Areas open to casual prospecting for visitors are described in detail. Simple as that.

Most of the prospecting sites included here are not promoted locally and are often unknown by local staff at visitor centers, museums, and even prospecting shops. This is the problem I set out to solve when I started writing. This book provides information on where to dig in Colorado for casual prospectors and visitors. If you learn about legal, open sites that are not in this book, please drop me an email at findinggoldincolorado@gmail.com and I will include it in my future writing with an acknowledgement to you for sharing the location.

> An important caution: There are NO other books available which work to avoid sending the reader to private land and are careful to guide prospectors away from areas subject to mining claim. Gold panning on private land or on someone else's claim is illegal and wrong. Even the other gold panning guidebooks fail on this key issue. This failing is the primary reason the *Finding Gold in Colorado* series exists!

There are several types of intended readers of this book:
- Tourists and locals who find themselves in one of the areas highlighted in the book. If you want to learn where to pan nearby, this book can help.
- People planning a trip to Colorado with a focus on gold prospecting and our gold rush history. When planning a trip, you can use this book, and its predecessor, alongside more conventional tourism resources. I have also included some tips on places to stay, and other basic logistics, here. I also frequently mention additional resources for each area

that are in my other book (*Finding Gold in Colorado: Prospector's Edition*) but this book still stands on its own.

- And, of course, Colorado gold prospectors like me who want to find more spots to dig. This book is for you in a way you will uniquely appreciate.

Whichever sort of reader you are, welcome! I hope you find this book helpful. I had a lot of fun researching and writing it and hope you have a lot of fun with it in hand!

I am also the author of the www.findinggoldincolorado.com website. A little bit of the information in my books is also on the website but there is much more that is unique. This book includes more and different dig sites, more driving tours, history, more everything. The structure of a book simply allows for better organization of this large collection of information. In fact, only two sites from this book are also on the website.

The website also has many things I didn't include in my books, such as equipment recommendations and prospecting techniques, links to some of the best YouTube videos on panning techniques, lists of all the prospecting clubs in the state, a consolidated list of all the prospecting stores in the state, and general educational information on gold deposits in Colorado. There are also lots of cool pictures online which would have made this book expensive and heavy if I included them here.

The website will also have updates to information provided here when that seems important. At least once every season, grab your pencil and copy any updates from the website into your copy of the book. That way you never need to buy a new edition of the book, you'll be current thanks to the website updates! I hope you check out the website regularly, I mean, hey, it's free right?

Note: To find the book updates mentioned above, go to www.findinggoldincolorado.com/wanderingprospector/ and read the comments below the article. All the crucial edits are provided there. That article will also tell you how to get more copies of the book and our other books too.

CAUTIONS AND DISCLAIMERS

ANTIQUITIES: The Federal Antiquities Act of 1906 (and later laws) protects artifacts and sites on public lands. This means removing anything of a historic nature from public lands is a crime. This includes removal of old mining artifacts, equipment, structures, or portions of structures situated on lands managed by the U.S. Federal government. Even old tin cans and other 'trash' can teach archeologists much about what happened historically. Do not take souvenirs from old mining sites, just pictures. The federal fines for this are quite large.

SAFETY: As they say in the mountaineers' books: Safety is an important concern in all outdoor activities. No guidebook can alert you to every hazard or anticipate the limitations of the reader. Therefore, the descriptions of roads, trails, routes, prospecting sites, and natural features in this book are not representations that any particular place or excursion is safe for you or your group. **When you use the information provided in this book, you assume responsibility for your own safety.** Under normal conditions, such excursions require the usual attention to road or trail traffic, road and trail conditions, weather, terrain, the capabilities of your group, and other factors. Keeping informed of current conditions, laws, regulations, and exercising thoughtful, attentive common sense are all key to a safe, enjoyable outing. The waterways carry pathogens of various sorts. Learn how to protect yourself from what is in untreated water. In some mining areas, the water may also carry dissolved metals which are detrimental to long term health if they are routinely absorbed into the skin. Use protective gloves and boots where needed.

Mercury is found in the paydirt in some waterways. It may be beads of free mercury which are visible or too small to see easily. It may also be a coating on some of the gold you find. Gold Prospecting is an excellent way to remove mercury from a waterway but that means it is your responsibility to handle it properly after it is out of the river. Whichever way it arrives in your concentrates, it is dangerous when it evaporates and is inhaled. Assume there is mercury or other toxins in your concentrates: store them in water or outside in a well-ventilated area to avoid accumulation of fumes. Store mercury and mercury coated gold separately in a water-filled container. A vial is good, then stored in a larger container also filled with water if you wish to be extra cautious. Get advice from your local prospecting club, store or online group on how to safely separate the mercury from your gold and how to recycle it properly. Don't let any of these cautions scare you off but don't take them lightly either. Mercury vapors can cause brain damage or death.

Keep in mind the risk of flash floods during the spring and summer when thunderstorms can drop a lot of rain into a drainage very quickly. In metro Denver, all the streets drain into the creeks and rivers. Some of those creeks rise 4-8 feet in an afternoon thunderstorm. This could kill you. Be smart if rain threatens upstream. Obviously, these risks can be amplified in the larger waterways like Cherry Creek, Clear Creek, and the South Platte River. Clear Creek, in metro Denver, kills people every year. Stay alert and know your fastest route to higher ground! In recent years, the city of Denver has added more exit points from the creek up to street level along Cherry Creek through central Denver. This was in response to a recent drowning death in a flash flood. These new ramps and ladders only help if you make a note of your nearest exit and stay alert.

Remember no gold is worth your life and it takes a lot of gold to pay for even the simplest visit to a hospital!

Safety First, Second AND Third!
...or as we used to say on the construction sites "Safety third!"

GENERAL PROSPECTING GUIDANCE

New to Gold Panning?
Start your adventure with a visit to a tourist site where they teach gold panning or join a prospecting club outing (a list of all the Colorado clubs is on my website findinggoldincolorado.com). For those near metro Denver, easy tourist choices include Vic's Gold Panning on CO-119 near Blackhawk, the Hidee Mine above Central City, the Argo Gold Mill in Idaho Springs, or the Phoenix Gold Mine just west of Idaho Springs and up the hill on Trail Creek Road.

Buy gold pans, a classifier, a suction pipette, and a plastic vial from your local prospecting store or via a link on my website findinggoldincolorado.com where there is a good article on recommended equipment including a starter kit.

I also strongly suggest new prospectors buy my other books, starting with *Finding Gold in Colorado: Prospector's Edition*. That book includes over 180 prospecting sites organized by area of the state and includes all of the major placer gold mining areas of the state as well as hard rock mining zones. This book is a sequel to that book. Trust me, you really want both!

Land Access:
The prospecting sites chosen for inclusion in this book were selected based on extensive research and often field-testing. These are all unclaimable locations where we don't have to worry about someone filing a federal mining claim and making us unwelcome at a site. It also means we do not have a right to use the land for prospecting however we see fit. Instead, we must follow the rules and regulations imposed by the governmental agencies responsible for the land or the private landowners themselves. Specifically, follow all posted rules, share the lands with other users, and defer to rangers, police officers, etc. The time to educate others is when they are interested, not while irate or confrontational. Arguing with a

ranger or law enforcement officer is unlikely to end well for the prospector. You may be able to educate a concerned ranger but avoid being argumentative. Instead, follow up later with them and their leadership if you feel the officer in the field was incorrect.

Many of the locations documented in this book are areas where the landowners and local governments have invested in infrastructure. Any prospecting you do should avoid damaging features built by the landowners such as roads, trails, dams, anti-erosion features, or creek-side park features. Most of the sites in this book allow pans and sluices; some allow electric power, while almost none allow gasoline-powered gear. A few restrict us to shovels and pans only. I have attempted to highlight the exceptions but rules change. It is always wise to ask first or to look for signage.

There are some real quirks such as:
- The Colorado State Parks System allows panning and 'informally' allows you to keep the flakes and fine gold despite it being technically state property. They explain this quirk openly on their beautiful and comprehensive website at http://cpw.state.co.us/thingstodo/Pages/Gold-Panning.aspx where they encourage people to try panning in several parks. Quite a few state parks with camping facilities are on gold bearing waterways so this permission is very helpful to traveling prospectors. One caution: some individual parks do not allow prospecting for various reasons. Always follow the instructions provided at the specific park you are visiting.
- The State Wildlife Areas (SWAs) are primarily set up for fishing and hunting. These areas are often on <u>private</u> lands with specific agreements allowing access for limited activity in limited seasons. Prospecting or any 'mineral investigation' is generally forbidden in SWAs except in a few state-owned areas - which I have documented in my books. In order to use SWA's, it is necessary to have a pass issued by Colorado Parks & Wildlife. They offer both day passes and annual passes (which include the state fishing license or a hunting license as valid options).

- Many local park districts allow non-motorized prospecting within the streambeds of their waterways, but some don't for various reasons. Please do not assume your permission to visit a park means you can 'go panning.' For example, in the City of Wheat Ridge, prospecting is allowed in the western part of the city on Clear Creek (Site A-02) but not in the rest of the city, despite having several creek-side parks, including one named Prospect Park. Similarly, Boulder County Open Space bans all prospecting (and flower picking, rock collecting, etc.) while some cities in Boulder County allow panning in the creeks in city owned/managed parks. Digging on dry land is almost always banned in municipal and state parks but Wheat Ridge allows digging in the banks and on dry land in a designated area on the north side of Clear Creek. The rules and land borders get tricky. Ignoring them can quickly mean we lose access to these lands so please be careful and respect the rules. My books are designed to solve this problem for you.
- Douglas County Open Space (DCOS) does not allow visitors to disturb or remove anything but much of the open space in northern Douglas County is not owned or managed by DCOS. This means there are plenty of spots available to dig in the south Denver suburbs of Lone Tree and Highlands Ranch despite those areas being part of Douglas County.
- Many small towns have no rules about prospecting whatsoever (other than a general ban on gasoline powered equipment of any sort in their parks). This is because they have not seen a need to restrict prospecting. Please help keep this true by being low-key while in their parks and by cleaning up after yourself and others.

By being low impact on the lands and other users, we can help maintain access for prospecting. History shows that annoying other users, or leaving visual impacts behind, will quickly trigger action from local governments and regulators. Taxpayers expect the local officials to protect the lands in their stewardship, so this often leads to a simple banning of all prospecting. Yes, I know I am being a bit repetitive here, but this is important! Again, be friendly and be careful!

My books are the ONLY books which only guide the reader to unclaimable Colorado lands open to the visiting prospector. Every other book includes quite a bit of private land where prospectors are trespassing. They also disregard restrictions in place on various state lands such as wildlife areas and private land with public access leases for hunting or fishing. Those other books also send prospectors to claimable public lands, much of which are actually part of active mining claims, making those visiting prospectors into thieves. Also, you can bet the first thing that happens when a book listing unclaimed, public lands gets published is that early readers of the book go claim that unclaimed ground. The book quickly becomes essentially useless. Don't be a trespasser and a thief - <u>put down the other books and use mine</u>!

<u>Navigation</u>:
Getting to a prospecting site is often most easily done by entering the GPS coordinates provided by this book into your GPS tool of choice, such as Google Maps. You will quickly have up-to-date driving instructions in hand. Just type the GPS coordinates in, in the same spot as you would a street address! Always type them as latitude, -longitude. The comma and the negative sign are important. In fact, if you leave the negative sign off, you will end up on a map of Mongolia – there IS gold in Mongolia, but this book isn't called Finding Gold in Mongolia!!

Learning to use an actual GPS or GPS app is a skill. Determining when you are between two established points is a matter of looking at how the Latitude and Longitude numbers at either end of a range differ and how your current position either is, or is not, between them. This is a skill best developed in an area you are comfortable navigating before you head out on a GPS guided adventure in new territory. GPS coordinates are used extensively in this book to guide you to staying where you are welcome. I myself use a GPS app on my smart phone. Whatever you use, handheld civilian GPS devices are only accurate to about 15 feet either way <u>at best</u>. As a result, I strongly suggest you stay at least 15 feet away from any unmarked boundary you identify using GPS coordinates, 30 feet is smarter. This way you can be confident you aren't an unintentional but unwelcome trespasser and a thief.

Within the description of a site, driving routes and sequences of GPS locations along waterways are generally listed from upstream to downstream as a matter of convention.

For each chapter, site numbers are also generally in order from upstream to downstream. Of course, there are exceptions because waterways meet at confluences and there are road access variations.

Mining Claims:

This book specifically does not discuss prospecting on public lands that are available to be acquired as mining claims. The rights of a claimholder give them the sole ownership of all valuable minerals on their claim. This ownership of minerals is just as real as your ownership of the flowers growing in your front yard. If you choose to prospect on claimable land, you must do your research beforehand to ensure you avoid any active mining claims. Additionally, while claims are initially marked at the corners and at the point of discovery in Colorado, there is NO requirement that these signs be maintained over time. Claim markers are often easy to miss in the field since claims are large and signs are small. Markers are also frequently vandalized and removed by other users of the public lands, or are destroyed over time by storms or other actions of nature. The lack of signage does not invalidate a claim. It is still theft to remove minerals from an active claim regardless of signage or other field evidence of the claim's existence.

The only way to avoid becoming a criminal is to do your research before going into the field to avoid the federal crime of mineral trespass or mineral theft. On the companion website to this book, I describe the basics of how to find and file a claim. The first part of this process is identical to what is needed to find public lands that are unclaimed and therefore open to digging. Refer to www.findinggoldincolorado.com for specifics in my articles on both finding your own mining claim and also how to buy a mining claim.

Environmental Impacts:
The hills around many mining towns were often rapidly denuded of trees during those first decades of mining. This was due to both logging and fires. To the casual observer, the land seems to have recovered well in the ensuing decades. Even so, in many areas of Colorado the results of this denuding can still be observed as large stands of pine, all of similar age.

While placer mining can be done with little or no environmental impact, the large-scale placer mining of the late 1800s and early 1900s did cause significant impacts in many cases. Hydraulic mining started in the 1870s and, in some cases, led to large amounts of sand and sediment being washed down into waterways with negative impacts on downstream fish habitat and agriculture. This led to the banning of hydraulic mining in Colorado via a state Supreme Court case in the early 20th century. The large dredges also caused problems as they dug up entire valleys. This process completely eradicated the existing riparian environment and left large cobble piles in its place. Again, to the casual observer, these areas may seem to have recovered as fish, mammals and various species of plants repopulated the areas over the ensuing decades. Often only an educated eye can still see the indications of prior mining activities but these areas are permanently transformed.

By contrast, small scale "artisanal" placer mining has been shown to be neutral or even beneficial to fish habitats as long as egg laying areas are avoided during that season. Responsible recreational prospectors also remove mercury, lead, and trash from the riparian environment. We are often cleaning up the impact of historic activity as well as more modern impacts such as trash washed into waterways from roads, lead from bird shot, and fishing tackle. The modest-sized prospecting holes left in the riverbed are often used by fish as a respite from the current. Generally, the powerful runoff from spring's high water is enough to eradicate anything a reader of this book would do in a streambed. Of course, this is less true for smaller creeks but, even there, one good gully-washer of a rainstorm will reset the creek bed material. Unfortunately, most members of the general public will not wait to see what happens to your holes, dams, etc. next spring. **Work to minimize or eliminate the visual impact of your visit before you leave your dig site.** This will avoid inspiring environment-related complaints from other users of the area.

Land Stewardship, Three Simple Rules:
1. Pick up any trash in your camping and digging areas. I do this at the beginning of my visit so I don't have to look at it while I'm on site. This also illustrates your approach to any other individuals who see what you are doing.
2. Avoid digging around woody plants such as trees and bushes. In high alpine environments of Colorado these plants generally take decades or centuries to mature. The pinch of gold caught in their roots is certainly worth less than those trees!
3. Fill your holes. This is especially important in areas used by non-prospectors or where it will be visible to a casual visitor. It is also important in virtually any dry land setting due to erosion concerns. Please take a few moments to shove the cobbles and loose tailings back in your hole.

Your Personal Responsibility:
Mistakes in the research and field-testing for this book have undoubtedly occurred despite my efforts toward accuracy. If you have any doubts, or if your presence and actions are challenged in the field, check the facts first. You are solely responsible for your actions. Regulations, rule interpretations and land status changes. Prospecting is a dynamic business!

Over the last five years, I visited and sampled many sites in this book. Some were tested for me by others. For other sites, I just confirmed the land status and didn't get to visit. For those, it'll be up to you to confirm safe access and the actual existence of gold too! Any discrepancies are my responsibility to correct so please do let me know of any changes in access. Email me at findinggoldincolorado@gmail.com or add a comment at www.findinggoldincolorado.com/wanderingprospector.

Testing a prospecting site for this book.

Flowers can be a lovely spring surprise at "Big Bend"

CHAPTER A: METRO DENVER & THE LOWER PLATTE RIVER

This chapter starts with a scattering of sites around metro Denver and then continues with the South Platte River starting at the west edge of Douglas & south edge of Jefferson Counties and running all the way to the far northeast corner of our state. Where it makes sense, sites from the other book are mentioned to highlight how the sites in each book fit in with each other in sequence from upstream to downstream. At virtually all of these sites, an electric highbanker will be easier to use than a stream sluice.

Site Number: A-22A-C
Site Name: Little Dry Creek

Little Dry Creek runs southeast through Westminster into Clear Creek. Sadly, the city of Westminster doesn't allow prospecting on their lands, but this lower stretch of creek is owned by Adams County Open Space, so we have easy access. For location context, the confluence of Little Dry Creek and Clear Creek is between sites A-16 and A-17 from the *Prospector's Edition*. Also, take note there is another "Little Dry Creek" in south metro Denver (mostly in Centennial), but that is an entirely different waterway. Amusingly, in the modern era at least, neither dry creek is actually dry!

Local Hints & Cautions:

- This area is prospected from time to time, but I have rarely seen other users here, other than those walking, running, or biking the greenway path. Do your best to stay out of their way. Access is easy thanks to the paved greenway path.

Gold Finding Tips:
- This is a fairly easy place to find a little color.
- There is some nice gold here, if you don't find it at first, keep prospecting!

Getting There: This stretch of the greenway is a bit west of Pecos St. from 64th Ave. north to above 68th Ave. The third access point has both the most parking and the longest walk. The rec path travels along the creek through this whole area.

A-22A - Canosa St: The South end of Canosa St at 67th Ave. at 39.8194, -105.0202; just park along the curb of this residential area and hike down to the creek.

A-22B - Dog Park: Zuni St. & 66th Ave. Use the provided parking lot and access the creek from either the north or south ends of the park.

A-22C - Adams County Open Space parking: On the south side of 64th Ave, just west of the bridge over the creek at 39.8124, -105.0165.

Locale: north metro Denver
Land Type: creek running through a suburban area
Land Manager: Adams County Open Space

Boundaries: From 39.8182, -105.0197 to 39.8153, -105.01464. Further upstream the creek is owned by Westminster; downstream it is private property.

Key Regulations:
- Read and follow all posted rules on park signage.
- All forms of prospecting are allowed within the banks of the river but no gasoline engines; there are good spots to sluice

or an electric high-banker will help immensely in other spots.

- Do not dig into the riverbank or on vegetated dry land or where landscaping has been done or undermine bridge supports, riprap (large rocks installed intentionally to control erosion, etc.) or other structures.

Nearby Attractions & Accommodations:
Other dig sites along Clear Creek.

Site Number: A-23A, B
Site Name: Clear Creek, Pecos to Broadway

This section of the greenway along Clear Creek is between Pecos Street and Broadway. This site is situated next to a neighborhood to the north and an industrial area on the south side of the creek. This is between sites A-16 and A-17.

Local Hints & Cautions:
- Both of the access options for this site involve a fair amount of walking. Pack your gear accordingly.

Gold Finding Tips:
- This would be a good place to use a hand pump like the Gold-N-Sand since there is bedrock here.
- There is some very nice gold here - if you don't find it at first, keep prospecting!

Getting There: This stretch of the greenway provides parking and access at a couple of points.

A-23A - Clear Creek Bottomlands: The best access point is at the intersection of 68th Ave. and Santa Fe on the south side of 68th. The address for this spot is 1001 W. 68th Ave. or 39.8200, -104.996. However, there is only parking there for 4-5 vehicles so you may find yourself parking in the neighborhood to the north or choosing to go further east to the next spot. By the way, a big thanks to Red Wilcox, inventor of the Gold-N-Sand hand pump and co-inventor of the Gold Cube for tipping me off to this site years ago. A bit of a walk but easy access.

A-23B - Twin Lakes Park: park at Twin Lakes Park on the south side of 70th Ave, just west of Broadway at 39.8234, -104.9903

Locale: north metro Denver
Land Type: creek running through a very urban area
Land Manager: Adams County Open Space

Boundaries: From 39.81611, -105.0016, downstream to Broadway Bridge where site A-17 (in the prior book of course) starts. Between the upstream boundary and the Pecos Street bridge to the west, the creek is on private property, owned by the industrial site to the south.

Key Regulations:
- Stay away from the designated wetlands area on the southside of the creek in the center of this property.
- Read and follow all posted rules on park signage.
- All forms of prospecting are allowed within the banks of the river but no gasoline engines; an electric high-banker will help immensely in some spots.
- This is a city park so fill any holes and be a 'good neighbor' to others using the park, especially those fishing and those with children. That said, I have rarely seen users here other than those walking, running, or biking the greenway path. Do your best to stay out of their way.
- Do not dig into the riverbank or on vegetated dry land or where landscaping has been done or undermine bridge supports, riprap (large rocks installed intentionally to control erosion, etc.) or other structures.

Nearby Attractions & Accommodations:
Other dig sites along Clear Creek.

Site Number: A-24
Site Name: Hoskinson Park

The previous book included a couple of Ralston Creek sites in Arvada. In the time since, I became curious to find the upstream edge of the decent gold in Ralston Creek. I knew that upstream of a certain point, there was almost no gold (approximately one tiny color per 3-4 pans of classified dirt), while further downstream there is good gold. This odd situation was very confusing to the original prospectors and wasn't explained for decades after the gold rush. It turns out the lower part of the creek has re-exposed an ancient gold bearing riverbed from 30-50+ million years ago. Not something an 1850s prospector would know! In fact, you could say this site is as much about where NOT to go, as it is where TO go!

Local Hints & Cautions:
- You will be noticed by locals going for walks. Be prepared to act as a friendly ambassador for our passion!
- Don't dig in developed park areas. This means staying away from the planted bank of the river in the parks. It also means avoiding any appearance of damage to erosion control work that has been done here.
- If you go outside of the area I have mentioned below, you will find heavily manicured areas where we are unwelcome (and there's no gold anyway due to the park construction).

Gold Finding Tips:
- Between Brad Martinez and I, we were able sample about 10 spots over the course of several miles west of Estes St (at 8800 west) and none of them had either large river cobbles or any gold to speak of. So, stay east of Estes Street to find gold in Ralston Creek.
- The gold is hiding where there are also larger rounded river cobbles since both were deposited by the ancient river.
- This park stretches quite a way along the creek from west to east but only the eastern part of the creek has gold. Seriously, don't even bother going west of Estes Street, there's no gold in that section of the park.

Getting There: From I-70 Exit 267 at Kipling St. head 1.3 miles north on Kipling to 58th Ave, then 0.6 miles on 58th to Garrison St. Take a quick left onto Garrison and then almost immediately right onto Brooks Drive. Brooks Dr. runs along the north edge of the park so just follow it to Estes St. or somewhat further east. Park either along the edge of the park or on a residential side street. **NOTE: This site is just west of Site A-18 in the original book.** Just west of site A-18 is Memorial Park which is also a good place to prospect, as long as you stay away from the manicured parts of the park. Very easy access from the parking to the creek.

Locale: Arvada
Land Type: small creek running through the minimally developed part of a city park.
Land Manager: City of Arvada, Parks Dept.

Key Regulations:
- Read and follow all posted rules on park signage.
- This is a city park environment so fill any holes and be a 'good neighbor' to others using the park, especially those walking their dogs and those with children.
- Do not dig into the riverbank or on vegetated dry land.
- Do not dig where landscaping has been done or undermine bridge supports, riprap (large rocks installed intentionally to control erosion, etc.) or other structures.

Site Number: A-25A-C **Site Name:** Goldsmith Gulch

Goldsmith Gulch runs through the Denver Tech Center on the south edge of the city of Denver and to the north, up through a residential area in southeast Denver. Along the way it crosses through some open space areas where there is good access and some gold. The gold is probably here because Goldsmith Gulch runs though the historic floodplain of Cherry Creek and re-exposes gold moved around by the larger Cherry Creek.

Local Hints & Cautions:

- You will be noticed by locals going for walks. Be prepared to act as a friendly ambassador for our passion!
- Don't dig in developed park areas. This means staying away from the planted bank of the river in the parks.
- If you go outside of the areas I have mentioned below, you will find heavily manicured areas where we are unwelcome (and there's no gold anyway due to relandscaping the creek itself during the park construction).

Gold Finding Tips:
- The gold is fine here but fairly easy to find as long as you stay away from the mucky, muddy areas. This is based on my experience at Hutchinson Park. I haven't sampled the other two parks.

Getting There: The parks provide parking and access at a several points:

A-25A: Rosamund Park with parking at 8051 E Quincy. This is just north of the I-225 DTC Boulevard freeway exit. Creek access is from Quincy Ave. north to Princeton.

A-25B: Hutchinson Park with parking at the Goldsmith Gulch Trailhead parking on the south side of E Cornell Ave, just west of S Tamarac Drive. This is just a bit north of Hampden Ave. Creek access is from here south to Eastman Ave. and north through **Bible Park** to Yale Ave. There is also extensive parking in Bible Park at 6802 Yale Ave. Very easy access

A-25C: Cook Park with parking at the Cook Park Recreation Center at 7100 Cherry Creek Drive South. Creek access is from East Mexico Ave. on the south to confluence with Cherry Creek on the north. NOTE: See also Site A-26 for commentary on digging Cherry Creek here.

Locale: Denver
Land Type: small creek running through a very urban area
Land Manager: City & County of Denver, Parks Dept.

Key Regulations:

- Read and follow all posted rules on park signage.
- This is a city park environment so fill any holes and be a 'good neighbor' to others using the park, especially those walking their dogs and those with children.
- Do not dig into the riverbank or on vegetated dry land.
- Do not dig where landscaping has been done or undermine bridge supports, riprap (large rocks installed intentionally to control erosion, etc.) or other structures.

Site Number: A-26A-F
Site Name: Cherry Creek in Denver

While almost all of Cherry Creek in the city of Denver is legal to dig, I have chosen to highlight areas with easy access, and where the creek remains fairly wild. These sites start just downstream (north) of Site A-20 in the prior book and continue downstream through the cities of Denver and Glendale to Site A-08. Moderately easy access in general, some slopes to handle.

Local Hints & Cautions:
- You will be noticed by locals going for walks. Be prepared to act as a friendly ambassador for our passion!
- Only dig downward in the bed of the creek.
- If you go outside of the areas I have mentioned below, be cautious about doing anything that might draw negative attention or involve trespassing.

Gold Finding Tips:
- The gold is fine here but fairly easy to find as long as you focus on areas where you see black sand or, even better, rounded river rocks at least as big as golf balls.
- Also look for areas where the creek has been running fast enough to strip away the top layer of sand in the creek bed, exposing the larger rocks. This is a case where digging in the faster current can be the most productive (rather than always focusing on inside bends) and sluiceable too.
- I have tested most but not all of these sites so have fun exploring the creek to find your honey hole. I have found spots along here that produce 8-12 colors per classified pan

so you can be sure the gold is out there waiting for you!

Getting There: The parks provide parking and access at a several points:

A-26A: Paul A. Hentzel Park with parking south from E Yale Ave. on S Elmira St. which ends in a cul-de-sac on the edge of the park at 39.6640, -104.8733; then just walk south into the park. Feel free to venture either upstream as far as the next bridge or downstream as far as you want...just avoid the golf course areas. There are some long walks here for full access.

A-26B: Cherry Creek Meadows with parking on Cherry Creek South Drive at the cul-de-sac at 39.6641, -104.8800 or along the road just before the cul-de-sac. Get here by going north on S Yosemite St. from Hampden Ave. While prospecting, stay out of the golf course property, that is Cherry Creek Country Club!

A-26C: Wabash Trailhead at 2596 S Wabash St. or 39.6711, -104.8895 then walk west across S Wabash St. into the park.

A-26D: Cook Park with parking at the Cook Park Rec Center, 7100 Cherry Creek Drive South; I suggest parking at the south end of the recreation center parking where it is a short walk to the river and it is easy to cross at the light. NOTE: See also Site A-25C for information on digging Goldsmith Gulch here.

A-26E: Cherry Creek Drive: From S Quebec St. all the way downstream to S University, Cherry Creek Drive is either following the north side of the creek or the south side of it. Use this road to choose your access points wherever you think it looks good.

A-26F: Speer Boulevard: From Downing Street downstream all the way into downtown Denver and the confluence with the South Platte River there is access along Speer Boulevard. Finding places to park is fairly easy for the upstream part of this and a bit trickier as you get into downtown Denver of course. However, channelized the creek may be here, I HAVE found gold even in this section. Access from the recreational path to the gold is easy along here.

Locale: Denver
Land Type: small creek running through a very urban area
Land Manager: City & County of Denver, Parks Dept.; City of Glendale

Key Regulations:
- Read and follow all posted rules on park signage.
- This is a city park environment so fill any holes and be a 'good neighbor' to others using the park, especially those walking their dogs and those with children.
- Do not dig into the riverbank or on vegetated dry land.
- Do not dig where landscaping has been done or undermine bridge supports, riprap (large rocks installed intentionally to control erosion, etc.) or other structures.

The South Platte River Upstream of Metro Denver

The next sites are on the South Platte River in the far southern part of Jefferson County and far southwest Douglas County. This area is between Salida, Deckers, Conifer, and Lake George broadly speaking. This area is not famous for its gold and isn't terribly productive. However, since it is a popular playground for many, it is of interest to prospectors who may find themselves in the area for other reasons (live there, fish, tube, picnic, and camp there for example).

Site Number: A-27
Site Name: Lone Rock Campground

Lone Rock USFS Campground is part of a fairly large area that is unclaimable and open to our recreational use.

Local Hints & Cautions:
- Be respectful of anglers you may encounter. This area is a popular fishing area. If you see someone fishing, give them a wide berth. Either set up downstream at least 50 feet or upstream even further. If they move toward you, there is no

need to move but wave rather than speaking if they are actively fishing.

Gold Finding Tips:
- The gold is very small through here so be prepared to catch and identify flour gold. I haven't sampled it myself yet.

Getting There: The campground is along Deckers Road aka CR-126 at 39.2533, -105.2363. You can also access the other side of the river along S Y Camp Road by heading south from the town of Deckers.

Boundaries: Upstream access starts at 39.2463, -105.2343 and downstream ends at about 39.2539, -105.2311; this means about 8/10 mile of river upstream of the campground and another 6/10 mile downstream!

Locale: southern Jefferson County just west of Deckers
Land Type: river in a narrow valley/canyon
Land Manager: USFS

Key Regulations:
- Respect posted rules in the campground.
- No gas-powered equipment near the campground.

Nearby Attractions & Accommodations:
Plan to stay at the campground perhaps. See recreation.gov for reservations, etc. Toilets on site. Lots of fly fishing here for trout.

Site Number: A-28A, B
Site Name: Bridge Crossing and Platte River CG

I combined these two into one site because they are close together and the access is continuous. Bridge Crossing Picnic Area is about 3.5 miles downstream from Lone Rock CG (the prior site, above). There are picnic tables, toilets, and parking. Good fishing and fun places to play in the water in summer including tubing. Access year-round but winter snow can create access issues. Platte River USFS CG is just a little further north. Have fun exploring this 1.5 miles of river access.

Local Hints & Cautions:
- If you have questions, you can stop in at Lone Rock CG to ask the CG Host since they also attend to these sites' visitors.

Gold Finding Tips:
- I haven't prospected here but there are some interesting inside bends and bedrock to explore.
- Expect the gold to be very fine.

Getting There: Both of these sites are on the east side of CR-67 aka Deckers Rd. between Deckers and Sedalia.

A-28A: Bridge Crossing Picnic Area USFS at 39.2913, -105.2071 and

A-28B: Platte River USFS CG (day use fee) at 39.2971, -105.2078.

Boundaries: Upstream access starts at the upstream edge of the picnic area at 39.2909, -105.2073 which is the road bridge and the downstream edge is 39.3012, -105.2011.

Locale: southwest of metro Denver between Sedalia and Deckers
Land Type: river running through a narrow valley
Land Manager: USFS

Key Regulations:
- Both sites have day use fees so bring small bills or a check book.

Nearby Attractions & Accommodations:
Camping on site! Fishing, swimming, tubing.

Site Number: A-29
Site Name: Ouzel CG

A fairly simple campground in a lovely spot, similar to its neighbors.

Local Hints & Cautions:
- Like the other sites along the river, this is a multi-use area so share respectfully with other campers, tubers and anglers.
- There is a campground host at Lone Rock Campground if you have questions.

Gold Finding Tips:
- Look for fine gold in the cobble bars along inside bends.
- The inside bend at the campground and the islands near it look interesting but I have not sampled this area.

Getting There: The Campground is at 39.3202, -105.1879 off of North Platte River Road and about 4 miles north of Deckers. From Sedalia, take CR-67 aka W Pine Creek Rd. southwest to North Platte River Road and head south.

Boundaries: Upstream access starts a LONG way upstream of the campground at 39.2128, -105.19216 with private property upstream of there. If you want to explore the upstream area, you may want to park at the dirt lot off the road at 39.3148, -105.1907. Downstream access is at the north edge of the campground at 39.3214, -105.1883.

Locale: southwest Douglas County, southwest of metro Denver
Land Type: river running through a narrow valley
Land Manager: USFS

Key Regulations:
- Follow posted rules in the campground. Day-use fee. Tents only for overnight.

Nearby Attractions & Accommodations:
Tent camp here, fish, tube, etc.

Site Number: A-30A-C
Site Name: Scraggy View-Osprey CG

With three developed areas and 2 ½ miles of river this next stretch provides many options for exploration.

Local Hints & Cautions:
- See comments on prior site.
- If you don't want to pay a day use fee, look for informal parking between the organized sites.

Gold Finding Tips:
- See prior site.

Getting There: From metro Denver, take US-85 south to CR-67. Take that to the left turn onto N Platte River Rd. The sites are along the road:

A-30A: Scraggy View USFS Picnic Area at 39.3280, -105.1874.

A-30B: Willow Bend USFS Picnic Area at 39.3429, -105.1789.

A-30C: Osprey USFS CG at 39.3494, -105.1766, on a great inside bend. No shade. Tents only. Toilets in summer only.

Boundaries: Upstream access starts near the upstream edge of the Scraggy View picnic area at 39.3269, -105.1872. The downstream edge is a bit past Osprey CG at 39.3509, -105.1746

Locale: Rural Douglas County, southwest of metro Denver
Land Type: river running through a narrow valley
Land Manager: Adams County

Key Regulations:
- Day use fees apply at the developed sites.

Nearby Attractions & Accommodations:
Camp at Osprey CG or one of the others a bit upstream.

Site Number: A-31A, B
Site Name: Nameless Accesses

With two separate pull-through parking areas between the road and the river, this section provides good access but no facilities at all. Both parking areas are on inside bends, so access is very easy here.

Local Hints & Cautions:
• See points included in the prior site.

Gold Finding Tips:
• See points on the prior sites.

Getting There: From Sedalia, take CR-67 southwest to the Platte River Road then turn right. This area is along CR-97, North Platte River Road.

A-31A: Park at 39.3751, -105.1697.

A-31B: Park at 39.3772, -105.1720.

Boundaries: Upstream access starts at 39.3725, -105.1672 and extends past both pull-off areas to the downstream point at 39.3802, -105.1701

Locale: rural southwest Douglas County, southwest of metro Denver
Land Type: river running through a narrow valley
Land Manager: USFS

Key Regulations:
• See prior site for details.

Nearby Attractions & Accommodations:
See prior sites.

Site Number: A-32
Site Name: Colorado Trail trailhead

The Colorado Trail, which runs from the southwest edge of metro Denver to the north edge of Durango comes through here. It crosses the road and river here at a convenient parking lot.

Local Hints & Cautions:
- See points included in the prior site.
- Stay away from the bridge while prospecting.

Gold Finding Tips:
- A foot bridge across the river takes us to the inside bend just upstream which looks appealing at this site.
- See tips on prior sites.

Getting There: Along the North Platte River Road north of the prior site at 39.4008, -105.1679. From Conifer take US-285 south to S Foxton Rd. (CR-97) and travel southeast to the parking lot. From Sedalia, take CR-67 southwest to the Platte River Road then turn right to follow the river road to the parking lot.

Boundaries: Upstream access is more limited on the east side of the river, beginning just upstream of the edge of the parking area while on the west side access is practically unlimited (and unfortunately fairly inaccessible before you get too far). Downstream access ends at 39.4006, -105.1687, which is just a bit downstream of the bridge.

Locale: Southwest of metro Denver
Land Type: river running through a narrow valley
Land Manager: USFS

Key Regulations:
- See prior site for details.

Nearby Attractions & Accommodations:
See prior site.

The South Platte River in Denver

The South Platte River in the actual city of Denver has surprisingly good gold in some spots due to a combination of sources. First, the obvious: fine gold washed down from the Alma & Fairplay area. Second: gold re-exposed from an ancient river that flowed about 30-50 million years ago, running from west Boulder County down to the front range near Coal Creek Canyon and then south along the front range through Denver. Some access sites were listed in my prior book. Here are the other major points along the river where there is convenient parking and good river access. You'll notice there are no GPS coordinates given for upstream and downstream boundaries here. That is because this entire stretch of river is owned by the county and open to our access as long as we stay off of private property up on the banks. Thanks to the rec paths access along the river is easy, it is often moderately tricky to get down the banks so watch your step. An electric highbanker is a great choice. There are areas a sluice will work but many others it would be hard to get setup properly.

Site Number: A-33
Site Name: South Platte at Publication Printers

There is a wide parking area on the river side of South Platte River drive across from Publication Printers. This area is a big inside bend so it is a tempting place to dig. This site is just a bit downstream of Site A-01 in my original book.

Local Hints & Cautions:
- You will be noticed by locals going for walks on the rec path. Be prepared to act as a friendly ambassador for our passion!
- Don't dig in developed park areas on the other side of the river...the gold is on this side anyway!
- This area is sometimes used by people living in their RVs and vans. Give them some space and lock your car.
- Access is moderately difficult here due to the steep banks.

Gold Finding Tips:
- The best cobbly part of the river bend is at the downstream end of the wide parking area.
- If you hit dense gray clay, that's the false bedrock. There is no gold in it so dig sideways!

Getting There: The parking area is between Evans Ave. and Jewell Ave. at 2055 S. Platte River Drive, 39.6796, -104.9998

Locale: Denver
Land Type: river
Land Manager: City & County of Denver, Parks Dept.

Key Regulations:
- Do not dig into the riverbank or on vegetated dry land.
- Do not dig where landscaping has been done or undermine bridge supports, riprap (large rocks installed intentionally to control erosion, etc.) or other structures.
- No gas-powered equipment. Everything else is good to go!

Site Number: A-34
Site Name: Overland Pond Park

Overland Park includes lots of interesting features related to the local ecology and gives us good access to the South Platte River just a bit downstream of Site A-01 in the *Prospector's Edition*. There is good, fairly easy access here to the river both upstream and downstream of the parking area.

Local Hints & Cautions:
- You will be noticed by locals going for walks. Be prepared to act as a friendly ambassador for our passion!
- Don't dig in developed park areas. This means staying away from the planted bank of the river in the parks.

Gold Finding Tips:
- Look for larger rock and cobble in the riverbed while avoiding sand bars that lack cobbles.
- Take advantage of the gentle inside bend near the parking

area.

- The area just downstream of the pedestrian bridge also has good cobble/gold deposits.
- If you hit dense gray clay, that's the false bedrock. There is no gold in it so dig sideways!

Getting There: The park provides parking and access just northeast of the intersection of South Platte River Drive and Florida Ave. at 39.6894, -104.9988.

Locale: Denver
Land Type: river
Land Manager: City & County of Denver, Parks Dept.

Key Regulations:
- Read and follow all posted rules on park signage.
- This is a city park environment so fill any holes and be a 'good neighbor' to others using the park, especially those walking their dogs and those with children.
- Do not dig into the riverbank or on vegetated dry land.
- Do not dig where landscaping has been done or undermine bridge supports, riprap (large rocks installed intentionally to control erosion, etc.) or other structures.
- No gas-powered equipment, everything else is fine.

The Mexican Diggings

Between this site and the next one is Johnson Habitat Park. My testing some years ago did not show gold in the river here. Feel free to give it a try if you like. I think the lack of gold has to do with the way this part of the river was manipulated over the years as the city developed. We do know that the river used to run further to the west, probably through the industrial area just west of the park. In fact, there is some interesting history here. The area near the west-center of the modern park at Jason St and Exposition Ave. was known as "The Mexican Diggings" during the early years of the Colorado Gold Rush. In 1857, a trapper named John Smith, who was married to an Arapahoe and living with the local Arapahoe Indians, guided a group of Mexican prospectors to this area. They spent enough time mining here that, in 1858, the Russell party found extensive evidence of their work on the east bank of the river.

Site Number: A-35
Site Name: Valverde Park

Valverde Park is across the street from the river but gives us a convenient place to park legally and access to some great areas to prospect. This site is a bit upstream of Site A-06 in the *Prospector's Edition* and happens to be one of the first places I explored on my own in central Denver. I'm really not sure why I left it out of the *Prospector's Edition*.

Local Hints & Cautions:
- You will be noticed by locals going for walks. Be prepared to act as a friendly ambassador for our passion!

Gold Finding Tips:
- Look for larger rock and cobble in the riverbed while avoiding sand bars that lack cobbles.
- The river just downstream of the Alameda Bridge is worth exploring but there are lots of other interesting spots in this stretch of urban river.
- If you hit dense gray clay, that's the false bedrock. There is no gold in it so dig sideways!

Getting There: The park provides parking and access just west of the river. It's on the west side of South Platte River Drive between Bayaud and Center (just north of Alameda Ave. at 39.7132, -105.0020.

Locale: Denver
Land Type: river
Land Manager: City & County of Denver, Parks Dept.

Key Regulations:
- Read and follow all posted rules on greenway signage.
- Do not dig into the riverbank or on vegetated dry land.

- Do not dig where landscaping has been done or undermine bridge supports, riprap (large rocks installed intentionally to control erosion, etc.) or other structures.
- No gas-powered gear, everything else is welcome.

Site Number: A-36
Site Name: Northside Park

Northside Park, aka Carpio Sanguinette Park, is just across the river from the western edge of the National Western Stock Show Complex. That means it is between Sites A-09 and A-10 in the first guidebook.

The first time I tried this site, years ago, I was at the NWSS Complex for the GPAA Gold & Treasure Show and I took the opportunity to sample a new spot. Despite high water conditions, it was very easy to find gold in my pan. It left me wondering how many show attendees realized they could be panning a few yards from where they parked for the show!?!

On the north edge of the City & County of Denver near I-70, this park used to be the Asarco Globe Smelter. In fact, it started operations in 1886 as the Holden Smelter. At its peak, the facility employed over 1,000 employees. Unfortunately, decades of pollution emissions from the plant led to serious environmental issues. When the plant was shut down, the property was turned into a park. Interpretive signs on site explain some of the history. Quite a few concrete structures remain from the old smelter's foundations, etc.

Local Hints & Cautions:
- Water levels and temperatures are fairly good here through the winter...this is a four-season site with many mild, sunny winter days and lots of heat in the summer. Sudden rains upstream can cause flooding here since there is no dam for a long distance upstream (at Chatfield State Park on the far south end of metro Denver). Be prepared for varying water levels.

- Don't dig in developed park areas. This means staying away from the planted bank of the river in the park itself.
- An electric high-banker or a gold pan are just about the only options here, I didn't see any good spots to set up a sluice, although that may be different when water levels are low in late summer and early autumn.

Gold Finding Tips:
- An electric high-banker, set up for very fine gold may be your best choice here if you want to do more than just pan.
- The South Platte River Trail runs along here providing access to as much river as you are willing to walk to.
- If you don't mind walking a short distance, upstream at 46th Ave. there may be some interesting gold since the I-70 bridges are right there. When the bridges were built, the construction bored down to bedrock and spewed all of the bored-out material into the river. So just downstream of the bridges there may be some good gold brought up from the holes. I haven't sampled here myself but in other, similar situations I have seen great results. You may be able to find parking just east of Washington on 46th Ave, which would put you right at the river. Have a look around on both sides of the river but avoid any "no parking" signs, etc.

Getting There: The park provides parking and access at a couple of points...
- 1400 53rd Ave., just west of Franklin St., then drive south and east through the parking lot to get close to the river.
- The point where 50th Ave. dead ends at the river at 39.7880, -104.9739; just be careful not to park in the actual cul-de-sac as it is needed by big rigs as a turn-around point after they make a delivery to the nearby business.

Locale: Denver
Land Type: river running through a very urban area
Land Manager: City & County of Denver, Parks Dept.

Key Regulations:
- Read and follow all posted rules on park signage.

- No gasoline engines but I can say an electric high-banker will help immensely.
- This is a city park so fill any holes and be a 'good neighbor' to others using the park, especially those fishing and those with children.
- Do not dig into the riverbank or on vegetated dry land.
- Do not dig where landscaping has been done or undermine bridge supports, riprap (large rocks installed intentionally to control erosion, etc.) or other structures.

Nearby Attractions & Accommodations:
The National Western Stock Show Complex is right across the river.

The South Platte River Downstream of Metro Denver

Downstream of metro Denver, the river is not particularly famous for its gold. However, it amuses me to think of all the greenhorn prospectors who could have been learning to pan while they walked along the South Platte from Nebraska through northeastern Colorado. They were "on their way to the gold" but also walking right past it. What a shame! Ah, well, more for us, right?

The gold here is very small but sometimes fairly plentiful. This is all flood gold that has washed downstream all the way from Denver (and upstream of there too of course!) Given the flood nature of the gold, some sites can be surprisingly productive while others are sparse and this can change from year to year. This part of the South Platte was significantly refreshed due to a major flood event on the river in September 2013. This so-called "100-year flood" made the news all over the world, caused at least 8 deaths and over $1 billion in damage, and yet drew cautious excitement from prospectors.

These sites extend downstream for over 100 miles, almost to the Colorado/Nebraska border.

If you do travel I-76, you may want to stop in Sterling to visit the **Overland Trail Museum** (*) which tells the migration story.

21053 County Road 26.5 (I-76 & US Hwy 6 East), Sterling, 80751; 970-522-3895.

Site Number: A-37
Site Name: South Platte River Trailhead Park

This first site is north of the city of Denver but still in the metro area. South Platte River Trailhead Park is popular with people walking the Platte River Trail and with anglers due to the large fishing lakes just south of the park. There are toilets available here. This access point is between sites A-10 and A-11 in the first guidebook. In fact, there is continuous prospecting access between A-10 and this section. Wow, that would be a bit of a long walk with prospecting gear!

Local Hints & Cautions:
• See points included in the prior site.

Gold Finding Tips:
• An electric high-banker, set up for very fine gold may be your best choice here if you want to do more than just pan.
• The South Platte River Trail runs along here providing access to as much river as you are willing to walk to, going south/upstream into Site A-10.
• Check out the long inside bend just upstream of the parking area.

Getting There: The park provides parking just off of Colorado Blvd., south of 88th Ave. Coming here via I-76, get off at the 88th Ave. exit and head west about a mile to the left turn onto Colorado Blvd. with the parking almost immediately on the left at 39.8548, -104.9340, or further south on the park road to more parking at 39.8524, -104.9393.

Boundaries: Go as far as you want upstream since this site connects with site A-10 in the first guidebook. Downstream to 39.8575, -104.9372 which is just past the 88th Ave. bridge. Then there is a short section of private property before we get to the next prospecting site.

Locale: Thornton, northeast metro Denver
Land Type: river running through an urban area
Land Manager: City of Thornton
Key Regulations:
- See prior site for details. Happily, the rules are the same for Denver and Adams County open space.

Nearby Attractions & Accommodations:
Large fishing ponds on this site and just north of 88th Ave. near the next site too.

Site Number: A-38
Site Name: Riverbend Park & Pelican Ponds Open Space

This large park and open space area is just north of the previous site with only a small piece of private property in between.

Local Hints & Cautions:
- See points included in the prior site.

Gold Finding Tips:
- The South Platte River Trail runs along here providing access to as much river as you are willing to walk going downstream toward 96th Ave.

Getting There: The park provides parking just off of Riverdale Rd., north of 88th Ave. Coming here via I-76, get off at the 88th Ave. exit and head west about a mile, to the right turn onto Riverdale Rd. with the parking on the right at 39.8631, -104.9343.

Boundaries: Upstream access starts just north of 88th Ave. at 39.8583, -104.9354 and continues on both sides of the river to 39.8745, -104.9073 which is north of 96th Ave.

Locale: Thornton, northeast metro Denver
Land Type: river running through a suburban area
Land Manager: Adams County

Key Regulations:
- See prior site for details.

Nearby Attractions & Accommodations:
Large fishing ponds on this site and just south of 88th Ave. too.

Site Number: A-39
Site Name: Elaine T Valente Open Space Park

This large park and open space area is just west of the South Platte River with almost a mile of river access.

Local Hints & Cautions:
- See points included in the prior site.

Gold Finding Tips:
- The South Platte River Trail runs along here providing access to lots of river exploration.
- There is good paydirt on a skim bar near the upstream end of the park.

Getting There: The park provides parking just off of 104th Ave., west of the South Platte river with parking at 39.8853, -104.9069.

Boundaries: Access on both sides of the river from 100th Ave. (no road access here) at 39.8781, -104.9041 northward to 39.8985, -104.8988 before running into private land.

Locale: Thornton, northeast metro Denver
Land Type: river running through a suburban area
Land Manager: Adams County & the city of Thornton

Key Regulations:
- See prior site for details.

Nearby Attractions & Accommodations:
Maybe you are in the mood to play some golf or disc golf at the

Adams County Fairgrounds just north of here (see next site).

Site Number: A-40A, B
Site Name: Riverdale Regional Park

This large park and open space area runs north along the river from 120th Ave. all the way to 136th Ave.

Local Hints & Cautions:
- Unfortunately for us, the river moves in and out of the eastern side of the park as it flows north. This means parts of the river appear to be part of the park but are actually on private land. Respect any fences and signage as you explore this section of the river.
- Exploring this site can mean a lot of walking so think about wearing good shoes and packing light!

Getting There and Gold Finding Hints:
A-40A: Riverdale Road & 120th Ave. – parking under 120th Ave. to access good gold. River access on the west bank from 39.9120, -104.8891 (a bit south of 120th Ave.) to 39.9138, -104.8796 (a little north of 120th). Then again on both sides of the river from 39.9150, -104.8756 to 39.9200, -104.8714 (which is well to the east of the point where the rec path bends left away from the river as it wraps around the lake).

A-40B: Parking on Henderson Rd. (aka 124th Ave.) about a mile west of US-85, at 39.9246, -104.8712 in a large dirt lot with a porta-potty. Walk a couple hundred yards east to the Platte River Rec path. There is a good paying bar just north of 124th and access to public land to the north and south on the west side of the river. The south edge here is 39.9217, -104.8702 with the north edge just north of the road bridge, at 29.9231, -104.8674. Unfortunately, a stretch of private land comes next but then public access continues to the north on both sides of the river from 39.9265, -104.8665 to 39.9323, -104.8620.

For access to the north end of the park, see the next site.

Locale: northeast metro Denver
Land Type: river running through a rural area
Land Manager: Adams County

Key Regulations:
- Electric pumps are allowed on high-bankers. No gasoline engines in the park.

Nearby Attractions & Accommodations:
The Adams County Fairgrounds are encompassed by this site which means this site has within it a spot from the first guidebook: A-12. The park here has many amenities, including golf and disc golf. Fishing is an option too.

Site Number: A-41
Site Name: Brighton Road Trailhead

When the E-470 Tollway was built, they created this E-470 Open Space. As a result, we gained this new access point to the rec path along the river.

Local Hints & Cautions:
- As with the prior site, this one is fairly large so be prepared for a fair amount of walking if you want to explore this site.
- The trailhead path from the parking lot to the river is about a third of a mile, followed by a walk along the paved rec path paralleling the river to whatever spot looks good, so bring wheels for your gear.

Gold Finding Tips:
- The South Platte River Trail runs along here providing access to lots of river exploration.

Getting There: The park provides parking just off of the intersection of Old Brighton Rd. and 136th Ave.

Boundaries: From upstream where the river enters Riverdale Regional Park by the pedestrian bridge over the river at 39.9415, -104.8608 to the trailhead path (just north/east of C-

470) and from the trailhead path continuing downstream to 39.9646, -104.8487. The river, both upstream and downstream of the area described, runs through private property.

Locale: northeast metro Denver
Land Type: river running through a rural area
Land Manager: Adams County

Key Regulations:
• See prior site for details.

Nearby Attractions & Accommodations:
See prior site for ideas.

Site Number: A-42
Site Name: Veterans Park & Colorado Park

The city of Brighton honors its veterans and celebrates the state of Colorado with these two adjacent parks. There is plentiful parking here close to the river, toilets, and a nice big inside bend to prospect.

Local Hints & Cautions:
• See points included in the prior site.

Gold Finding Tips:
• This is a tricky place to find gold. Look for streaks of black sand in the bed of the river or places larger gravel and rocks could hide gold. My friend Paul Schiffer and I were able to find fine gold in both of those ways. Paul says to be sure to sample the south end of Veterans Park!

Getting There: 405 Bridge St, Brighton 80601; on the west edge of town and just a few blocks west of the US-85 & CO-7 interchange along CO-7 aka 160th Ave. The river is west of the parking lot so take a left as you enter the lot and park closer to the river if possible.

Boundaries: Legal access from upstream at 39.9841, -104.8340 (south of there the rec path moves away from the river because the river is private property), north under the road, to the park and beyond to 39.9905, -104.8293 which is reached by following the dirt trail along the east side of the river. The legal access ends where the river is owned by the sand and gravel operation on the west side of the river.

Locale: Brighton, northeast of metro Denver
Land Type: river running along the edge of the town
Land Manager: City of Brighton

Key Regulations:
• Follow all posted signage.

Nearby Attractions & Accommodations:
Explore the town of Brighton. It has managed to maintain a cool small-town vibe despite being fairly close to population centers on the front range.

Site Number: A-43
Site Name: Pearson Park in Fort Lupton

In 1838, well before the gold rush, Lieutenant Lancaster Lupton built a trading fort here on Adobe Creek. The town developed later when agriculture became a focus of the area and now has about 8,000 residents.

This large park and open space area has baseball and soccer fields as well as fenced dog parks and walking trails. Oh, and good river access of course!

Local Hints & Cautions:
• Use the pedestrian bridge to get to either side of the river.
• If you used a cart and are heading downstream, you will most likely want to stay on the dirt path on the west side of the river.

Gold Finding Tips:
• Another tricky site but there is definitely fine gold here. As

at other South Platte River sites, look both for larger cobble where the gold can hide, and black sand streaks in the river bed.
- Notice the large inside bend and island at the downstream end of the park.

Getting There: Pearson Park is on the north side of CO-52 & just west of US-85. Leave your vehicle in the parking lot for the baseball diamonds at 40.0812, -104,8247. Walk east to the river on the path around to the south side of the diamonds.

Boundaries: Legal access is from the Highway 52 bridge north along and beyond the dirt path to 40.0909, -104.8183; note: the undeveloped land south of CO-52 is private.

Locale: Fort Lupton, northeast of metro Denver
Land Type: river running through a suburban area
Land Manager: Fort Lupton and CDOT

Key Regulations:
- Follow all posted signage.

Nearby Attractions & Accommodations:
Just north of the prospecting site, take time to check out the South Platte Valley Historical Park. It features the actual fort (rebuilt but including original adobe bricks, etc.) and other historical highlights. If you need a place to stay, the Motel 6 Fort Lupton is on the east side of the river just south of CO-52. It's basically walking distance from the dig area.

Site Number: A-44
Site Name: Riverside Park in Evans

The little town of Evans is a suburb of Greeley on the southeast edge of Greeley and right on the South Platte River. When Greeley was originally organized (1870) it was a sort of commune with bans on all temptations such as gambling, prostitution, and of course alcohol. Evans promptly sprung up 'next door" to provide all of those services. More recently, when cannabis was legalized in Colorado, Greeley banned pot sales so Evans happily stepped up again to host the new cannabis shops.

Local Hints & Cautions:
- Don't dig in developed park areas. This means staying away from the planted bank of the river in the park itself. It also means avoiding the swimming beach area.
- An electric high-banker or a gold pan are the best options most of the time here. Setting up a sluice may be possible if water levels are right.

Gold Finding Tips:
- The gold is very fine here. However, this area is rarely prospected so look for the obvious collection spots.
- The upstream and downstream edges of this area are reported to be the most productive.

Getting There: The park is basically at the intersection of CO-85 and the South Platte River. The address is 4000 Riverside Parkway, Evans CO. Parking is just off of 42nd St. at 40.3703, -104.6842; or coming in from the north, go past the big parking lot for the baseball fields to the smaller lot at 40.3722, -104.6821.

Boundaries: From 40.3660, -104.6959 at the US-85 bridge to 40.3745, 104.6778 after which there is private property.

Locale: Evans
Land Type: river running alongside a city park
Land Manager: City of Evans, Parks Dept.

Key Regulations:
- Read and follow all posted rules on park signage.
- All forms of prospecting are allowed within the banks of the river but no gasoline engines; an electric high-banker will help immensely.
- This is a city park so fill any holes and be a 'good neighbor' to others using the park, especially those fishing, playing on the beach, and those with children.

- Do not dig into the riverbank or on vegetated dry land.
- Do not dig where landscaping has been done or undermine bridge supports, riprap (large rocks installed intentionally to control erosion, etc.) or other structures.

Nearby Attractions & Accommodations:
Support the local businesses of your choice in Evans.

Site Number: A-45A, B
Site Name: Rainbow Bridge

Rainbow Bridge runs over the South Platte River in central Fort Morgan. The bridge marks the western edge of Riverside Park, a large multi-use park which is the crowning jewel of the Fort Morgan Parks Department. Fort Morgan is in far northeast Colorado, just about an hour from the Nebraska border. I-76 runs through town on its way along the river from metro-Denver to Nebraska.

The first permanent settlement was started here in 1865. Initially named Camp Wardell, it was set up to protect travelers and supplies headed west to the gold mines of Colorado as the was one of the main routes. It was renamed Fort Morgan the next year in honor of Colonel Christopher Morgan who had died in the line of duty that first year. The fort closed in 1868 and the area was abandoned. In 1884, Abner Baker from Greeley organized the town we see today, just to the south of the old fort ruins. Today there are about 11-12 thousand residents.

This site has been popular with Gold Cube owners who have been impressed with how much gold they can catch with their Gold Cubes here. You know what they say, "Cube it or lose it!" A big thanks to Red Wilcox and Mike Pung (co-inventors of the Gold Cube) who first turned me on to this site back in 2013 or 2014.

Part of what is cool about this site is how far into the northeast plains of Colorado we are here. We are closer to Nebraska and

Kansas than we are to Denver or the Rockies and yet there is good gold to be had!

Local Hints & Cautions:
- Water levels and temperatures are fairly good here through the winter...this is a four-season site with many mild, sunny winter days and lots of heat in the summer. Sudden rains upstream can cause flooding here since there is no dam for a long way upstream. Be prepared for varying water levels.
- Don't dig in developed park areas. This means staying away from the south banks of the river along the park.

Gold Finding Tips:
- The gold here is very small but fairly plentiful. However, it is also very patchy. Look for black sand, that's where the gold will be. This is probably flood gold that has washed downstream all the way from Denver (and upstream of there of course!) Given the flood nature of the gold, this site can be surprisingly productive. In fact, that has left some of us wondering about other, more ancient, sources for this gold. The area was significantly refreshed due to a major flood event on the South Platte River in September 2013.
- An electric high-banker, set up for very fine gold may be your best choice here.
- If you are sluicing, keep in mind that water levels can change quickly, blowing out a sluice if it has been raining upstream. Folks with riffle sluices should consider doing a quick clean up each time they are about to walk away from their sluice.

Getting There:
A-45A: From I-76 Exit 80, turn north onto highway 52. The developed park is almost immediately on your right. If you cross the bridge, you already missed the developed park but not your parking! Parking location: There is parking just east of the bridge on the north side of the river. This is the area for prospectors to park and head out to dig. There is also a toilet here. In theory you could park in the developed city park, but it is a longer walk to the river and you are more likely to draw

unwanted attention and concern. Please don't.

A-45B: This site is at the east edge of Rainbow Park Open Space and is little used. From I-76 Exit 82, turn north on Barlow Road. Drive to the end of the road where it forks, just past the Silver Spur RV Park. Take the left fork into the parking area at 40.2676 -103.7750 and walk north to the river. This area is day use only, no camping, no fires.

The walking paths through the natural part of the park do connect these two sites but since these two sites are almost two miles apart, having two access points is nice. Yes, you read that right, there is almost two miles of river to explore here!

Boundaries: From the bridge, downstream through the park to GPS coordinates 40.268781 -103.772719. Both sides of the river are city property as well as the riverbed. Use the longitude number on your GPS reading to gauge the boundary since this site runs from west to east.

Locale: City of Fort Morgan
Land Type: river running through an urban area
Land Manager: City of Fort Morgan, Parks Dept.

The land upstream of the bridge is owned by a local concrete company and the Western Sugar Cooperative, both private properties, so don't trespass. Downstream of our eastern boundary is all private ranchland.

Key Regulations:
- Read and follow all posted rules on park signage.
- All forms of prospecting are allowed within the banks of the river but no gasoline engines; an electric high banker can help you avoid carrying buckets of paydirt around although sluicing spots can be found seasonally.
- This is a city park so fill any holes and be a 'good neighbor' to others using the park, especially those fishing and those with children.
- Do not dig into the riverbank or on vegetated dry land.

- Do not dig where landscaping has been done or undermine bridge supports, riprap (large rocks installed intentionally to control erosion, etc.) or other structures.

Nearby Attractions & Accommodations:
The south edge of the city park (which itself is along the south edge of the river here) includes an area for 10-12 RVs to park overnight for free. Electrical service is provided but no other hookups. There is a playground, restrooms, and picnic shelters right next to the RV area. Remember that alcohol and cannabis consumption is not allowed in the park. At the west edge of the RV area, there is a pedestrian underpass to get to the south side of I-76 and then to the many shops and restaurants of downtown Fort Morgan. Don't miss the Dairy Queen, it's in easy walking distance. Explore on foot and enjoy!

Site Number: A-46A-E
Site Name: City of Sterling Open Space

As mentioned above, Sterling is the home of an interesting museum about the overland migrations that created the state of Colorado. The city, and this prospecting site, are about 45 miles downstream of the site in Fort Morgan. This large park and open space area is along the South Platte River with almost a mile of river access.

Local Hints & Cautions:
- This site is quite undeveloped, but if you do run into curious locals, please be a friendly ambassador for our gold prospecting passion.

Gold Finding Tips:
- This is a sandy, braided section of the river with multiple islands and gravel bars to explore. This braided condition also makes choosing where to sample more difficult. A tight grid pattern may be the best sampling bet.
- Look for black sand streaks that have been created naturally by the river, larger rocks that could create a collection point for traveling gold flecks and also the high

points of sandbars where the river drops material as it flows past plants.

Getting There:
A-46A: The upstream access point is right at the south boundary of the city property at 40.6112, 103.1965 on the west side of the river. To get here, from I-76, Exit 125, take Chestnut St. west to Edith Ave. Follow that south to Iris Dr. and carefully turn left, following the signage down the dirt road to the parking area near the river.

The next access points are just west of the I-76 interchange at Exit 125, on the north side of Chestnut St. (business loop I-76). There are access points for this open space on both sides of the river:

A-46B: Overland Trail Recreation Area: On the east side of the river, just west of the Holiday Inn Express, there is a turn-off into a fairly large dirt lot. From there, follow the trails west or north to the river.

A-46C: North Riverview Rd.: driving north from Chestnut, on Riverview past the Rest Area, look for dirt parking and paths west to the river. There are several including across from the CDOT yard and behind the golf course maintenance buildings.

A-46D: Follow Riverview Rd. north past the golf course, just past the railroad tracks to the left turn at 40.6323, -103.1758, then go a few feet, past the fence, turn left again and follow the dirt drive to a dirt parking lot with signage near the river.

A-46E: Once over the river bridge on Chestnut, turn right onto Harris St. and follow it to the end. Take the walking path east to the river.

Boundaries: As mentioned above, the upstream boundary of the city property is at 40.6112, 103.1965 The downstream limit is quite a walk north of access points B and C, past the golf course, the railroad and all the way to 40.6441, -103.1784.
Note: due to the north-south flow of the river here, it's

sufficient to keep an eye on your latitude as you explore.

Locale: City of Sterling on the northeastern plains of Colorado
Land Type: river running across the plains
Land Manager: City of Sterling

Key Regulations:
* Damming up a stream is forbidden.
* Dogs are welcome, even off leash.
* Read and respect any rules posted onsite.

Nearby Attractions & Accommodations:
The Overland History Museum is the other logical stop while here. There is also a nice rest area near site A. The town has quite a few charming older brick public buildings and is worth exploring as well.

Site Number: A-47
Site Name: Sedgwick

Tiny Sedgwick, population less than 200, was named for Fort Sedgwick. The fort was named for Major General John Sedgwick who fought with the Union Army during the US Civil War. The actual fort wasn't here, it was on the prairie between here and Julesburg. The fort was established in 1864 and operated until 1871. U.S. militia guarded the stage route along the South Platte and the telegraph line. In retribution for the Sand Creek Massacre, Cheyenne and Sioux men attacked and looted the fort in January 1865. Records from the time highlight the mixed activity of the soldiers, sometimes maintaining calm and other times instigating violence with the local American Indians. There were also many soldiers who deserted to go west to the gold mining areas in search of their fortunes. The fort was never well constructed and was abandoned in 1871. To learn more about this history, visit the Fort Sedgwick Museum in Julesburg.

Local Hints & Cautions:
* This is a very small patch of land. Be respectful of the

private property upstream and the state game reserve downstream.

Gold Finding Tips:
• See prior site for tips.

Getting There: Take Exit 165 off of I-76 and head north past Lucy's Place restaurant and toward Sedgwick. Pull off on the side of the road, either just before or just after the bridge over the river, about 1.5 miles north of the interstate highway.

Boundaries: Legal access is limited to the riverbed on either side of the bridge. From the edge of the bridge access upstream extends 200 feet while it reaches just 50 feet downstream.

Locale: the south edge of Sedgwick on the northeastern plains
Land Type: braided river flowing across a plain
Land Manager: CDOT

Key Regulations:
• No gas-powered prospecting gear within 50 feet of the bridge.

Nearby Attractions & Accommodations:
Try a meal at Lucy's Place, it is well loved by locals and visitors. Quite affordable as well.

Site Number: A-48
Site Name: Ovid

Ovid is another tiny little outpost on the plains with around 300 people living there. It was named for a founding resident, Newton Ovid.

Local Hints & Cautions:
• Park on the north side of the River Road to stay off of private property. Do NOT block the road, it is used by the locals.

Gold Finding Tips:
• See prior site tips.

Getting There: A large chunk of the river right on the south edge of town here is owned by the county and there is easy access via "River Road". From I-76, Exit 172, head north toward Ovid but turn onto River Road before the bridge over the river. The turn is at 40.9495, -102.3876. Park anywhere that appeals to you along the road before it turns south at about the ¾ mile mark. Park on the north side of the road to stay off of private property.

Boundaries: Legal access starts at the bridge, extending downstream for 4.5 miles to 40.9476, -102.3038. Sadly, there is no other legal access along the way downstream unless you ask a local rancher for permission.

Locale: Ovid, northeast plains of Colorado
Land Type: braided river running across the plains
Land Manager: Sedgwick County

Key Regulations:
• Follow any posted signage.

Nearby Attractions & Accommodations:
Stop in at Luna's Bar & Grill (217 Main St.) for a burger and a beer or some tasty fried rocky mountain oysters! I hear the tacos are great too.

Site Number: A-49
Site Name: Julesburg

The farthest northeast town in Colorado, Julesburg is just a couple of ranches away from the Nebraska border. With a population of over 1,300 people, it's the largest town in the area. It was named for the Jules Beni's trading post originally on this site and was on the Pony Express route from Missouri to California. Beni was also a horse thief and when captured, was tied to a fence post, and shot multiple times. Isn't it odd that they didn't rename the town?!?

Local Hints & Cautions:
• Park in such a way as to stay out of the way of vehicles towing trailers who may want to turn around.

Gold Finding Tips:
• See prior site tips.

Getting There: From I-76 Exit 180, head north on US-385 over the bridge to the right turn into a dirt parking area at 40.9743, -102.2519. Park and walk east or southeast to the river. It may also be possible to drive toward the river on the dirt road to a parking spot closer to the river. Be aware of varying water levels if you choose to park near the river! Alternatively, turn right on CR-32.5 just before the bridge and park off of the road wherever there is safe space on the shoulder between the bridge and 40.9752, -102.2484.

Boundaries: Legal access is from US-385 downstream as long as you are south of 40.9754 Latitude. Along CR-32.5 this is about 2/10 of a mile of river. The legal area forms a triangle bounded by US-385, CR-32.5 and the latitude mentioned above.

Locale: South edge of Julesburg in the northeastern corner of the state
Land Type: braided river running across the plains
Land Manager: Sedgwick County

Key Regulations:
• Follow all posted signage.
• No gas-powered prospecting equipment within 50 feet of US-385.

Nearby Attractions & Accommodations:
The Fort Sedgwick Museum is in town at 114 E 1st St. See the history of the fort in the Sedgwick site description. There is also a rest stop/visitor center just off the expressway at 20934 CR-28, 40.9664, -102.2514 with some interesting displays about the American Indians, the Pony Express, and the bison.

CHAPTER B: BOULDER COUNTY

<u>First a couple of history tourism nuggets:</u>
The quirky little Gold Hill Cemetery is south of town at 1170 Dixon Hill Rd., about ½ mile out of the town of Gold Hill on the right side of the road after the turn off for Gold Run St.

The town of Salina in Four Mile Canyon was settled in 1874 as hard rock mining started to boom in the area. Its population peaked at over 350 in around 1900 but today is less than 100. To get there, drive about four miles up Four Mile Canyon Road from CO-119 to the area around the turnoff to Gold Run Road. The crossroads is at 40.0505, -105.3525 in western Boulder County.

The area was first prospected in 1859 of course and a small placer mining camp was established in 1864 but the placer deposits were sporadic at best. The real boom started in 1874 when a group of miners from Salina, Kansas found gold-tellurium ore and put up the first buildings including a post office. A year later, Camp Salina had a hotel, miners boarding houses, a blacksmith, a church, a couple general stores and even a telegraph line down to Boulder. The church has its own funny story: Henry Myring, who discovered the successful Melvina Mine built the church and the blue house next to it. His wife Sarah insisted on the church being built because she was so upset with the low morals in town! It was used as a mine assay office in the 1920s and 1930s so I suppose it either worked so well at improving local morals that it wasn't needed any more or it failed miserably. In any case, the church has been restored and the Myring family descendants still live in the area! To see the little church, head up Gold Run Road about ½ mile and look on the right. It is used as an Episcopal Church and local community center.

Multiple significant hard rock mines operated in the area for decades including the Melvina, the Richmond, the Railroad, and the Mayflower. The town had three different stamp mills over the years to serve the mines. By 1925, the post office closed since rising costs had closed the mines and only about 25 people remained in town. However, the town never quite died so there is still something to see.

Prospecting Opportunities

As I mentioned in my prior book, neither the City of Boulder nor Boulder County allow any prospecting on the land they own. This limits our opportunities to find unclaimable prospecting areas in the area. In the first guidebook, I included a few good sites. This book has many more! Truth in advertising compels me to warn my readers that many of these sites have only small amounts of very fine gold, but others will provide happy surprises.

Site Number: B-04A, B
Site Name: Middle St. Vrain Creek

This area presents an interesting access challenge. Just like in Boulder Canyon (see my prior guidebook for legal sites), and at the next site (just below), we have patches of access due to a how two different facts fit together. First, we aren't allowed to prospect on Boulder County land which means when the creek is too close to the county-owned road, we can't touch it. Second, if the creek is within 200 feet of the center line of this road, it is unclaimable. This was done to preserve scenic values on the Peak-to-Peak Highway when it was originally federally funded. So, we are looking for USFS lands where the creek is within 200 feet of the road but not so close that it is on the county's lands. Got all that? Good! I used the Boulder County website to confirm exact property boundaries and here's what I found...

Local Hints & Cautions:
- Do not dig into the bank next to the road (you shouldn't be that close anyway, but I have to say it) or interfere with any erosion control structures. This area is prone to flooding during rain events so the authorities are very sensitive about this issue.

Gold Finding Tips:
- I haven't tested this spot myself yet, but there are a couple nice river bends including the inside bend at the parking.

Getting There:
B-04A: This spot is along CO-7 aka South St. Vrain Dr. west of Lyons. Pull off at parking at 40.1388, -105.4864.

B-04B: This spot is, of course, also along CO-7 aka South St. Vrain Dr. west of Lyons. Pull off of the highway at Riverside Dr. (40.1453, -105.4724) and follow that to parking on the right about 200 yards along at 40.1459, -105.4706.

Boundaries:
B-04A: The creek at the parking is open to casual use. The upstream limit here is at 40.1373, -105.4906 and the downstream limit is 40.1401, -105.4846. Upstream and downstream beyond these limits is private property.

B-04B: The creek at the parking is open to casual use. The upstream limit here is at 40.1451, -105.4713 and the downstream limit is 40.1472, -105.4689. Upstream of this section of creek is Boulder County property where no one is allowed to remove anything, not a pebble or a blade of grass; downstream is private property.

Locale: In the foothills southwest of Lyons
Land Type: creek on the edge of a residential area
Land Manager: USFS

Key Regulations:

- None noted. Ask the local USFS ranger station before using gas-powered equipment; they are likely to have restrictions since the creek is small and close to the road.

Nearby Attractions & Accommodations:
There is a lot to explore in the mountains west of here including other prospecting sites in Chapter B of *Finding Gold in Colorado: Prospector's Edition.*

Site Number: B-05
Site Name: Middle St. Vrain Creek

This area is open to casual use for the same reason as the prior site. Take a moment to read that info to understand the situation here.

Local Hints & Cautions:
- Do not dig into the bank next to the road (you shouldn't be that close anyway, but I have to say it) or interfere with any erosion control structures. This area is prone to flooding during rain events so the authorities are very sensitive about this issue.

Gold Finding Tips:
- I haven't tested this spot yet myself but there are a couple nice river bends including the inside bend at the parking.

Getting There: This spot is along CO-7 aka South St. Vrain Dr. west of Lyons. The obvious parking spot is a dirt loop off of the road at 40.1710, -105.4097.

Boundaries: There is about 700 feet of creek with a gentle inside bend and good parking. The upstream limit here is at the parking, specifically 40.1711, -105.4109 and the downstream limit is 40.1707, -105.4084. Either side of this section of creek is Boulder County property where no one is allowed to remove anything, not a pebble or a blade of grass.

Locale: In the foothills west of Lyons

Land Type: creek in a hilly area
Land Manager: USFS

Key Regulations:
- None noted. Ask the local USFS ranger station before using gas-powered equipment; they are likely to have restrictions since the creek is small and close to the road.

Nearby Attractions & Accommodations:
See prior site.

Site Number: B-06A, B
Site Name: South St. Vrain Creek

This area presents the same interesting access challenge as the prior sites. Take a moment to read the introduction to first site in this chapter to understand why we are legal to dig here.

Local Hints & Cautions:
- Do not dig into the bank next to the road (you shouldn't be that close anyway, but I have to say it) or interfere with any erosion control structures. This area is prone to flooding during rain events so the authorities are very sensitive about this issue.

Gold Finding Tips:
- I haven't tested this spot yet myself but there are a couple nice inside bends including the inside bend at the parking.

Getting There:
This spot is along CO-7 aka South St. Vrain Dr. west of Lyons. The obvious parking spot is a dirt loop off of the road at 40.1682, -105.3812

Boundaries:
B-06A: This spot is about 800 feet of creek with an inside bend and good parking. The upstream limit here is at 40.1670, -105.3831 and the downstream limit is 40.1684, -105.3808. Either side of this section of creek is Boulder County property

where no one is allowed to remove anything, not a pebble or a blade of grass.

B-06B: Just downstream, this section is bigger, about 1/3 mile and also has a nice inside bend with a good parking. The upstream edge (with a good dirt parking loop) is at 40.1718, -105.3682; there is another well used, wide creek-side parking area near the downstream edge, which is at 40.1716, -105.3629.

Locale: In the foothills west of Lyons
Land Type: creek in a hilly area
Land Manager: USFS

Key Regulations:
- None noted. Ask the local USFS ranger station before using gas-powered equipment; they are likely to have restrictions since the creek is small and close to the road.

Nearby Attractions & Accommodations:
See prior site.

| Site Number: B-07 |
| Site Name: Left Hand Canyon Drive |

A quarter mile section of Left Hand Creek is open to us thanks to the federal government setting it aside for a reservoir which hasn't been built. NOTE: THE STATUS OF THIS SITE IS DISPUTED AS OF 2025. THERE IS A CLAIM WHICH MAY, OR MAY NOT, BE VALID > check signage before digging.

The road runs right along the creek here making access easy. This spot is just below the confluence of Left Hand Creek and James Creek, so we are potentially catching gold from both sources.

Local Hints & Cautions:
- Do not dig into the bank next to the road or interfere with any erosion control structures. This area is prone to flooding during rain events, so this is a very sensitive issue.

Gold Finding Tips:
- I haven't tested this spot yet myself but there are a couple nice inside bends to sample.

Getting There: Follow Left Hand Canyon Drive west into the hills to parking at 40.1079, -105.3365 where there is a nice pull-off area on the creek side of the road.

Boundaries: The upstream limit here is 40.1052, -105.3365 and the downstream limit is 40.1084, -105.3360. Either side of this section of creek is City of Boulder property where no one is allowed to remove anything, not a pebble or a blade of grass.

Locale: In the foothills east of Jamestown
Land Type: creek in a hilly area
Land Manager: USFS

Key Regulations:
- None noted. Ask the local USFS ranger station before using gas-powered equipment; they are likely to have restrictions since the creek is small and close to the road.

Nearby Attractions & Accommodations:
Hop west on the same road to explore little Jamestown (and one of the other prospecting sites in Chapter B of *Finding Gold in Colorado: Prospector's Edition*).

| Site Number: B-08 |
| Site Name: Bohn Park |

As South Saint Vrain Creek traverses Lyons, it passes through Bohn Park. With multiple easy access points there is plenty to explore here. The park also offers a wide range of features from sports fields to playgrounds, picnic areas, and pavilions.

Local Hints & Cautions:
- Do not dig into the bank next to the park or interfere with any erosion control structures. This area is prone to

flooding during rain events so the authorities are very sensitive about this issue.
- If other park users are curious, be a friendly ambassador.

Gold Finding Tips:
- I haven't tested this spot yet myself but there are a couple nice inside bends.

Getting There:
The official entrance to Bohn Park is on the west side of Lyons at 199 2nd Ave. In that case, take the first right as you enter the park and leave your vehicle in the first lot on the right at 40.2202, -105.2663 and walk north to the creek.
However, additional access points include:
- The north end of CR-69 at 40.2182, -105.2724, then walk north to the creek.
- Prospect St. & 4th Ave. Park at the curb and enter the park at 40.2204, -105.2700 and walk south to the creek.
- Park St. east of 2nd Ave. park near 40.2214, -105.2648 and walk south to the park.

Boundaries:
The upstream limit here is next to the CR-69 access at 40.2183, -105.2725 with access to both sides of the creek down to the confluence with North Saint Vrain Creek (feel free to try this area too). From there access continues on the park side of the creek past the pedestrian bridge to 40.2203, -105.2629 where the creek enters private property. Downstream of this point there is a mix of private and town land but the creek is so manipulated and channelized it isn't worth the trouble.

Locale: in the Town of Lyons, north of Boulder and west of Longmont
Land Type: city park
Land Manager: Town of Lyons

Key Regulations:
- Pans only unless you show your other gear to a ranger and get permission to use it. They MIGHT give permission to use a stream sluice.

- Only dig downward into the riverbed.

Nearby Attractions & Accommodations:
Explore the charming town of Lyons.

Prospecting in Longmont
The next few sites are all on City of Longmont property.
Overall, the rules are:
- Only dig downward in the bed of a stream, NO digging on dry land or where there is vegetation.
- Pans only, no fancy equipment is allowed.
- Only dig in natural areas (not where landscaping reaches the edge of the creek).
- Stay out of wetlands areas.
- Leave the area exactly as you found it, no piles, no holes
- If a ranger or police officer challenges you, be agreeable, they are just trying to do their job; the rules are somewhat unclear, and individuals may have their own interpretation.

Site Number: B-09A, B
Site Name: St. Vrain Greenway.

This section of the St. Vrain Greenway varies between lovely at the upstream end and semi-industrial at the downstream. However, the rec path makes it easy to move around.

Local Hints & Cautions:
- Parts of the St. Vrain Creek Greenway are owned and managed differently than others. The section included here is all City of Longmont. Please don't go out of bounds.

Gold Finding Tips:
- The gold here is fine. Sampling where there are exposed cobbles in the water is a good start.

Getting There:
B-09A: Rogers Grove Park: Turn off of Hover at 220 Hover, 40.1637, -105.1305 (there is private property upstream of this

point even if it does look like part of the park).

B-09B: Izaak Walton Nature Area: Turn off of Sunset at 18 S. Sunset St. 40.1620, -105.1212; if you park here, remember walk upstream to the west side of Sunset St before digging.

Boundaries: According to the city rules and regulations "The St. Vrain Creek in Longmont is open to all uses from the western City limits (at Airport Road) to Boston Avenue." And explicitly closed to all access for the section after that. I suggest something more restrictive to avoid areas where human manipulation of the creek and land ownership makes our activity questionable. Please only pan between Rogers Grove Park, on the creek at 40.1651, -105.1272, and S. Sunset St. at 40.1611, -105.1214.

Locale: Longmont
Land Type: creek in a park and through a mixed-use commercial area.
Land Manager: City of Longmont

Key Regulations:
• See the list and the beginning of this section.

Nearby Attractions & Accommodations:
The town of Longmont has lots to offer, check out the cool downtown along Main St. or some of the local craft breweries, there are many! A fun note: Left Hand Brewing's tap room is at the intersection of Boston and St. Vrain Creek, just downstream from this panning area; I highly recommend an after-digging toast there for those who enjoy great craft beer. They often have beers on tap which aren't available elsewhere!

Site Number: B-10
Site Name: Left Hand Creek Park

Left Hand Creek is named for Chief Niwot of the Arapaho Nation. The English translation of Niwot is "Left Hand". Chief Niwot was the local chief of the Arapahoe who were in this area when the gold rush started. He consistently advocated for peaceful relations but also told the early prospectors they were not welcome to settle in the Boulder Valley. Those prospectors moved up into the mountains to places like Gold Hill but sadly others did not respect his wishes in the same way.

The park offers nice amenities including a play structure, picnic areas with barbeque grills, and toilets.

Local Hints & Cautions:
• None noted.

Gold Finding Tips:
• The gold here is fine. Sampling where there are exposed cobbles in the water is a good start.
• The upstream part of this section is more wild and more likely to be productive.

Getting There: The main parking area is at 1800 Creekside Dr., Longmont at the park. An alternative is to park on the street just east of Hover/N 95th St. on Corporate Center Circle where there is dirt parking alongside the road at 40.1353, - 105.1293; then walk south to the creek across the field. Another option is to leave your car at Kanemoto Park at 1151 S Pratt Pkwy toward the downstream end of this section.

Boundaries: Upstream access starts at Hover St. and downstream access ends at Main St., east of the park quite a way. Upstream of Hover is Boulder County property where no panning is allowed.

Locale: Longmont
Land Type: creek in a residential area and along a park
Land Manager: City of Longmont

Key Regulations:
• See the rules at the top of this section.

Nearby Attractions & Accommodations:
The town of Longmont has lots to offer, check out the cool

downtown along Main St or a local craft brewery.

Site Number: B-11
Site Name: Left Hand Creek at Main St.

This parking lot behind a big box store provides access to Left Hand Creek through a natural section of the waterway that hasn't been "manipulated" by humans.

Local Hints & Cautions:
• Stay away from all structures and erosion control work.

Gold Finding Tips:
• See prior site.

Getting There: The parking is just west off of Martin St., just south of Ken Pratt Drive at 40.1508, -105.0936

Boundaries: Upstream access starts at 40.1487, -105.1001 and downstream access meets the next site listed below. Upstream of the limit is private property so please don't pan there even though it "looks" ok.

Locale: Longmont
Land Type: creek running near a shopping area
Land Manager: City of Longmont

Key Regulations:
• See the list at the beginning of this section.

Nearby Attractions & Accommodations:
See prior site.

Site Number: B-12
Site Name: St. Vrain Open Space Access

This large parking lot provides access to the St. Vrain Creek from its confluence with Left Hand Creek downstream through a natural section of the waterway that hasn't been "manipulated" by humans. Other amenities include restrooms and a sheltered picnic table area.

Local Hints & Cautions:
- Stay away from all structures and erosion control work.

Gold Finding Tips:
- The gold here is fine. Sampling where there are exposed cobbles in the water is a good start.

Getting There: The park provides parking just off of 119th St., just south of Ken Pratt Drive at 40.1541, -105.0743

Boundaries: Upstream access starts just upstream of the confluence with Left Hand Creek, at 40.1552, -105.0883 and downstream access ends at the 119th St. bridge. Don't go upstream of the point where the creek stops being natural.

Locale: Longmont
Land Type: creek running near an old industrial area
Land Manager: City of Longmont

Key Regulations:
- See the rules at the beginning of this section.

Site Number: B-13A, B
Site Name: St. Vrain Greenway & Sandstone Ranch Park

With almost 3 ½ miles of river and a well-developed park, this area is a whole destination. Sandstone Ranch is a very large park with lots of amenities such as sports fields, toilets, picnic shelters, and lots of parking lots. It also provides access to a huge piece of City of Longmont property that is barely mentioned as part of their open space system at this point. St. Vrain Creek flows through this area so we have miles of creek to explore...if we have the inclination to walk that far!

Local Hints & Cautions:

- Stay away from all structures and erosion control work.

Gold Finding Tips:
- The gold here is fine. Sampling where there are exposed cobbles in the water is a good start.

Getting There:
B-13A: St. Vrain Greenway: This is at the upstream edge of this area at 10190 CR-1 (40.1485, -105.0552) with signage, paved parking, etc. Walk north to the river.

B-13B: Sanderson Ranch: The park is just south of Ken Pratt Drive. Turn south onto Sandstone Dr. at 40.1604, -105.0367 and follow that road to its south end where there is parking at the ranch house. This is the closest parking to the open space and the creek. From here, walk the trails south to the bridge over the creek. Access extends upstream along for about 1.2 miles and downstream for 2.2 miles of waterway.

Boundaries: Upstream access starts at CR-1 and downstream access ends at the CO-119/Ken Pratt Blvd. bridge.

Locale: Longmont
Land Type: creek running across the rural plains
Land Manager: City of Longmont

Key Regulations:
- See the Longmont rules at the beginning of this section.
- There may be signage warning visitors to stay on the trails due to ongoing reclamation. Please respect this.

Site Number: B-14
Site Name: Gross Reservoir

The city of Denver owns the large Gross Reservoir here on South Boulder Creek. Despite the odd name, this is a very lovely area. (The reservoir is named after a Colorado water engineer.) South Boulder Creek below the dam remains on Denver property for a distance offering us prospecting access. If you have a copy of *Finding Gold in Colorado: Prospector's Edition*, you know there is a site (C-01) on this creek further upstream. The gold I have seen there is why I was excited to find these additional sites.

Local Hints & Cautions:
- This area is active with anglers and kayakers so share with them and stay out of their way if they are on site when you are looking to set up.
- The hillside down to this spot is very steep. Use Caution!
- Water levels in the creek can vary quickly due to dam releases. Use Caution!

Gold Finding Tips:
- Gold has been moving into this area from the mountains to the west for millennia so don't worry about the dam "stopping the gold", it's here already.
- In the creek, look for large boulders with crevices that can trap gold.

Getting There: This site is in the foothills west of Rocky Flats National Wildlife Refuge. Take CO-72/Coal Creek Canyon Rd. west from CO-93 to Crescent Park Dr., follow that to the right turn onto Gross Dam Rd. The parking is just east off of Gross Dam Rd. at 4368 Gross Dam Rd., 80302

Boundaries: Upstream access starts at the access road at 39.9393, -105.3501 and downstream access continues to 39.9382, -105.3457. Upstream of the limit is the fenced off dam operations area. Downstream is Boulder County land, so it is off-limits. Note: Only prospect from the parking lot side of the creek and into the creek. Rangers will issue trespass tickets to those on the other bank of the creek so stay in the water.

Locale: in the foothills west of Boulder
Land Type: creek running in a rural area
Land Manager: Denver Water

Key Regulations:
- None noted. Follow all posted signage.

Nearby Attractions & Accommodations:
Try some fishing or hiking perhaps?

Site Number: B-15A-C
Site Name: Johnson Gulch Reservoir

The city of Denver owns a little reservoir here. Two stretches of South Boulder Creek below the dam are on Denver property for a distance, each offering us prospecting access. This site is just west of Eldorado Canyon State Park. The last part of the driving route follows the creek with the road next to the legal prospecting zone.

Local Hints & Cautions:
• THERE MAY BE NO PRACTICAL ACCESS TO THIS SITE since Kneale Rd is generally gated off within the park. In theory, hiking in may be possible but has not been confirmed.

Gold Finding Tips:
• Gold has been moving into this area from the mountains to the west for millennia so don't worry about the dam "stopping the gold", it's here already.

Getting There: This site is in the foothills southwest of Boulder. Take CO-170/Kneale Rd. southwest from CO-93 through Eldorado Springs and Eldorado Canyon State Park, all the way to CO-170/Kneale Rd (IF it is open to public access). Look for parking that makes sense along the side of the road in the prospecting area below or park at the very end of the road in a way that will not block others (39.9331, -105.3084).

Boundaries:
B-15A: Upstream access starts just below the dam at 39.9331, -105.3096, around a tight curve and downstream access continues to 39.9318, -105.3070. Upstream of the limit is the dam operations area so please don't pan there even though it "looks" ok. Downstream is Boulder County land so it is off-

limits.

B-15B: After transiting a patch of Boulder County land, the creek moves back north onto Denver property from 39.9328, -105.3049 downstream to 39.9354, -105.3069

B-15C: An additional spot in the reservoir's claim withdrawal area on BLM land, along Kneale Rd. from 39.9316, -105.3014 to 39.9305, -105.3001 for about 315 feet of access wedged between two private homes. Please be very respectful of the homeowners. Parking on the west side of the road at 39.9315, -105.3009.

Locale: in the foothills west of Boulder
Land Type: creek running in a rural area
Land Manager: Denver Water, BLM

Key Regulations:
- Do not use gas-powered equipment at this site (this is due to being so close to the road).
- Do not dig in a way that could compromise the road, avoid all erosion control rock work.

Site Number: B-16
Site Name: Eldorado Canyon State Park

South Boulder Creek runs through Eldorado Canyon State Park just before reaching 'civilization" in the form of the town of Eldorado Springs. Like many state parks, gold panning is allowed as a recreational activity with some strict restrictions.

Local Hints & Cautions:
- This area is active with anglers and kayakers so share with them and stay out of their way if they are on site when you are looking to set up. Remember, they tend to move on fairly quickly, so they won't likely be in your way for long. There are rock climbers too - share the space.

Gold Finding Tips:

- In the creek, look for large boulders with crevices that can trap gold.
- The gold I found here at the east edge of the park was very fine. Come prepared with a good classifier, especially since you won't be bringing concentrates home with you (see the rules below).

Getting There: This site is in the foothills southwest of Boulder, just downhill from the prior site about 1/3 mile or so. Take CO-170/Kneale Rd southwest from CO-93 through Eldorado Springs to Eldorado Canyon State Park. The park offers many spots to pull off along the road up the canyon.

Boundaries: South Boulder Creek runs through the park for almost a mile, from 39.9303, -105.2953 to downstream at 39.9324, -105.2808.

Locale: in the foothills west of Eldorado Springs
Land Type: creek running in a canyon
Land Manager: Colorado State Parks

Key Regulations:

- The rules in a state park are very strict in the spirit of resource protection while encouraging outdoor recreation. Pans and shovels only. Restore your dig area to original condition by filling holes and smoothing out tailings piles. No removal of material is allowed, so don't bring your concentrates home with you. Finally, if you find anything particularly interesting, like a nugget or artifact, the parks ask you to take it to the visitor center to show a ranger. Maybe you will get famous, and they will display your discovery behind glass in the visitor center! Everything in the park belongs to the state, keep that in mind.

Nearby Attractions & Accommodations:
Check out the visitor center to learn more about the park. Try some fishing, climbing, or hiking perhaps?

CHAPTER C: CENTRAL CITY & GILPIN COUNTY

A Little Historic Tourism

Central City has not one, but six historic cemeteries. One is simply the municipal cemetery, the others were all organized by local fraternal organizations like the Masons, International order of Odd Fellows, Knights of Pythias, Red Men, and the Catholics. The Masonic cemetery isn't open to visitors, but the others are and happily they are all clumped together in a high valley above town at 39.8093, -105.5300. There are some very, very old graves here so history buffs will enjoy a visit. The setting is also lovely with summer wildflowers and aspen groves in the cemetery grounds. In the municipal cemetery see if you can find the grave of Dennison J. Ely. He died here in 1898, having been one of the pioneering prospectors in 1858.

Prospecting Opportunities

The new dig sites in this chapter are all right around the town of Blackhawk. Both the town and the county are very prospector friendly. In fact, the first time I visited one of these sites, a couple city workers stopped to chat with me for a minute, just to be friendly and welcoming. Note: All three of the following sites take some scrambling to get down to the creek and back.

Site Number: C-07
Site Name: Eureka Creek at Gregory Diggings

This tiny piece of Gregory Gulch/Eureka Creek is at the site of John Hamilton Gregory's big discovery. He followed the placer gold in the streams from Arapahoe City up to this point where he looked up at the hillside on the south side of the gulch and recognized the hard rock gold ore there. Eureka! The first large scale hard rock gold deposit in the state was discovered. Happily, today the town of Blackhawk owns this little piece of the seasonal creek upstream of where it has been undergrounded in the name of progress.

Local Hints & Cautions:

- Water levels in the creek are variable. You can count on water during the spring melt but at other times this site may be entirely dry. Often insufficient water for sluices.

Gold Finding Tips:

- The gold here is a mix of very fine gold and chunkier stuff.

Getting There:
From Central City, drive downhill on Gregory Street to the GPS coordinates 39.8006, -105.5002, turn into the paved parking lot on the left/north side of the road. The dig site is on the upstream edge of the parking lot. Park on the very uphill edge of the lot where you will be out of the way of any city workers who need to use the lot.

Boundaries: The upstream boundary between private land and this site is at the bridge at 39.8005, -105.5004. The downstream boundary is at the drain grate on the edge of the parking lot. This is only about 60 feet of creek bed!

Locale: Gilpin County on the border of Central City and Blackhawk.
Land Type: Tiny, narrow creek in a developed area.
Land Manager: City of Blackhawk

Key Regulations:

- Do not let any tailings get into the city drain grate at the downstream end of this site.
- Do not do any damage to the banks which might cause erosion or lead to damage to the road.
- Do not block the movement of city vehicles when you use the parking lot.

Nearby Attractions & Accommodations:
The Gilpin History Museum, historic Central City, the Hidee Mine, and many other attractions, both historic and dig sites. See *Finding Gold in Colorado: Prospector's Edition* for details.

North Clear Creek upstream of Eureka Creek
This park of North Clear Creek has much less gold so it is easy to see why Gregory turned left up the smaller Eureka Creek. Even so, if you'd like to see what's upstream yourself, here are a couple sections of creek you can explore with your gold pan.

Site Number: C-08
Site Name: Upstream edge of Blackhawk

This stretch of North Clear Creek feels a bit wilder since it is all the way up at the north end of town, away from the buildings. The creek actually runs along Apex Valley Road for most of this section.

Local Hints & Cautions:
- Water levels in the creek are variable but rarely extreme except for a couple weeks in the spring. Very sluiceable.

Gold Finding Tips:
- The gold here is all very fine and relatively sparse from what I could see. There are some decent sources of gold further upstream (on private property) so it's possible you will find larger gold and prove me wrong!

Getting There: From Blackhawk, take CO-119 uphill to the north to Apex Valley Road. Take a left onto Apex Valley Rd. and find safe, roadside parking where it is legal.

Boundaries: The upstream boundary between private land and this site is at 39.8200, -105.5179. The downstream boundary is around the curve of CO-119, east of the intersection with Apex Road. The exact boundary is 39.8191, -105.5107

Locale: Gilpin County on the border of Blackhawk.
Land Type: tiny, narrow creek in an undeveloped area
Land Manager: City of Blackhawk & CDOT

Key Regulations:
- Do not do any damage to the banks which might cause erosion or lead to damage to the road.
- Do not block the movement of other vehicles when you park.

Nearby Attractions & Accommodations:
See previous site.

Site Number: C-09
Site Name: North Clear Creek just above Blackhawk

This stretch of North Clear Creek is just a little bit downstream of the prior site. It includes a long inside bend that may provide some interesting spots to pan depending on water levels.

Local Hints & Cautions:
- Water levels in the creek are variable but rarely extreme except for a couple weeks in the spring. Very sluiceable.

Gold Finding Tips:
- As with the prior site, the gold here is all very fine and relatively sparse from what I could see. There are some decent sources of gold further upstream (on private property) so it's possible you will find larger gold and prove me wrong!

Getting There: From Blackhawk, take CO-119 uphill to the north to the GPS coordinates and find safe, roadside parking where it is legal. Since parking is a challenge, it may be wiser to park at the post office, or park somewhere easy closer to town and walk upstream to this area.

Boundaries: The upstream boundary between private land and this site is at 39.8179, -105.5072. The downstream boundary is around the curve of CO-119, east of the intersection with Apex Road. The exact boundary is 39.8072, -105.4975.

Locale: Gilpin County in Blackhawk north of most of town.
Land Type: tiny, narrow creek in an undeveloped area
Land Manager: City of Blackhawk & CDOT

Key Regulations:
- Do not do any damage to the banks which might cause erosion or lead to damage to the road.
- Do not block the movement of other vehicles when you park.

Nearby Attractions & Accommodations:
See prior site.

CHAPTER D: IDAHO SPRINGS & CLEAR CREEK COUNTY

Santiago Mill Driving Tour

From Georgetown, head southwest on FR-248.1 to FR-248.2D and continue to 39.6431, -105.7708. Remains of an ore bin from 1911, a mill from 1935 and a large water tank from 1948 remain on site. The water tank was said to hold 3,700 gallons when operational.

Prospecting Opportunities

I only have a couple of sites to add here because I was fairly thorough about land access last time. I am excited to add the second site because it is accessible in spring when Clear Creek is too big and fast to be safe.

Site Number: D-08A-C
Site Name: Clear Creek at Dumont

This access to Clear Creek next to the little town of Dumont is mostly interesting because the town of Dumont was established during the initial gold rush. However, it's tough to see that important history in what little remains of the town now that I-70 has cut through it.

Local Hints & Cautions:
- Water levels in Clear Creek are variable based on spring melt and rainstorms upstream.

Gold Finding Tips:
- The gold here tends to be quite small but fairly easy to find.
- Densely packed material with a bit of muddy clay content mixed into the gravel and cobble seemed to pay the best (as is often the case!) if you can find it.

Getting There: This site on Clear Creek is in the long gap between sites D-02 and D-03 in the first guidebook. Take Exit 234 from I-70 and then choose one of the following destinations.

D-08A: Head west on CR-308 on the north side of the interstate, until it passes under I-70 to parking next to the bus stop at 39.76660, -105.6268. From there walk east, across the street, to the old trail along the creek.

D-08B: Get off of I-70 at Exit 234 and head to the truck weigh station on the south side of the road. Instead of turning left toward the truck scales, turn right into a small parking area next to the river. DO NOT block the flow of truck traffic or you will quickly get ticketed and towed. To access this area from the south side of the creek, either
- from D-08A or D-08C, follow surface streets to 39.7644, -105.6086 where there is parking on the north side of West Dumont Rd.
- Or, follow the instructions below to get on West Dumont Road and follow it west to the county provided parking at 39.7644, -105.6086.

D-08C: Get off of I-70 at the Stanley Rd exit (#235) and head to the south side of the interchange and take the right onto the dirt parking lot or cross the creek and turn right immediately on to West Dumont Rd., another immediate right puts you in the public parking lot.

Locale: Rural Clear Creek County next to I-70.
Land Type: medium sized creek running through a valley

Land Manager: part Clear Creek County, part CDOT

Boundaries:
D-08A: The upstream part of this site is all CDOT land stretching from the parking area downstream to 39.7652, -105.6151 where C-08B starts.

D-08B: starts where C-08A ends and continues as county land (pans only) to 39.7649, -105.6089. This section is the most natural looking part of the creek.

D-08C: The north half of the creek starting at 39.7643, -05.6120 downstream to the end of C-08B and then both sides of the creek all the way past the next I-70 exit to 39.7640, -105.5928.

Key Regulations:
• Avoid any interference with the flow of trucks to the scales.
• Pans only on Clear Creek County lands, no such restrictions on CDOT land but gas-powered equipment is always banned within 50 feet of CDOT structures.
• Avoid any interference with anti-erosion rock work.

Nearby Attractions & Accommodations:
Hmm, a Taco Bell, I bet that would make a great reward for my wife if she comes along when I want to prospect here!

Site Number: D-09
Site Name: Trail Creek downstream of the Phoenix Mine

Trail Creek runs from the famous LaMartine mining district, downhill past the Phoenix Mine (which has an excellent tour and a decent gold panning area in their section of the creek) to Clear Creek next to I-70. Along the way most of the creek is on private property but there is also a public access section fairly close to the bottom of the hill. This area is just southwest of Idaho Springs.

Local Hints & Cautions:

- Be sure to park FULLY off of the road so others can get by on this narrow, dirt, mountain road.
- Easy access at the upstream end, a bit of a scramble at other spots.

Gold Finding Tips:
- The gold here tends to be quite small and can be challenging to find.
- The creek falls rapidly downhill through this section so it is tricky to find places where the gold accumulates. Stop and look at a section to imagine where the stream widens out and slows down when it is flowing heavily.
- Check multiple spots, the gold is patchy.

Getting There: Trail Creek Road is fairly simple to get to from I-70. Get off at exit 239, and take Stanley Road under the freeway and west to Trail Creek road. Follow it uphill to the small parking spot (fits just a couple vehicles), at 39.7498, -105.5562 which isn't very far uphill at all. It's right next to a point where the creek runs through a culvert under the road.

Locale: Rural Clear Creek County along a dirt road but near I-70. Accessible via any car.
Land Type: small sized creek running through a valley
Land Manager: Clear Creek County

Boundaries: from 39.7501, -105.5570 downhill to 39.7503, -105.5541 which is just a stone's throw uphill from Stanley Rd.

Key Regulations:
- Avoid any interference with the flow of vehicles on the road.
- Pans only on Clear Creek County lands.

Nearby Attractions & Accommodations:
The Phoenix Mine which has tours, gold panning lessons, a bit of prospecting gear for sale or rent(!), etc. Idaho Springs is also just downhill offering food and other prospecting locations as described in the *Prospector's Edition* book.

CHAPTER E: SUMMIT COUNTY & THE BLUE RIVER

Peru Creek Driving/Hiking Tour

The Peru Creek basin above Keystone is famous for its "multi-metal" mines. The mines in this area had a type of ore containing commercial quantities of gold, silver, copper, lead, and zinc. The largest and most famous was the Pennsylvania Mine which was opened up in 1879 by J.M. Hall. It was a multi-metal mine typical of the area, producing all five of the metals I just mentioned. In 1893, the mine had its biggest year, shipping 14 million pounds of ore to the mill. The mine continued regular operations until 1908, and then opened and closed several times until it finally closed forever in the mid-1940s. It was famous for its production but, unfortunately, also for the water pollution it created. Remediation efforts by the U.S. EPA in 2006 addressed the issue with the mine being plugged in 2014/15. Until then, the Snake River was considered the most impaired river in the state of Colorado, and the Pennsylvania Mine was the major contributor to the problem. Even today, Peru Creek is bright white but that is due to natural sources of minerals according to the Colorado Division of Reclamation, Mining and Safety. By hiking upper Peru Creek, it is fairly easy to find one of the springs which produces the naturally white tinted water.

This is a driving and hiking tour because while the biggest

sites are drive-up locations, some of the smaller mines are hike-in only. The hikes are very rewarding with lots of wildflowers and dramatic views of the valley and surrounding mountains. The hikes are of moderate difficulty because while the walking is on old mine roads for the most part, this is an extremely high elevation area with most mines above 11,000 feet elevation.

To start the tour, head out Montezuma Road from Keystone. As you travel southerly, you will get to Peru Creek Road (FR-260) on the left at about the 4-mile mark on your odometer (39.5920 -105.8710). From here it is another 4 miles up a bumpy dirt road to the turnoff for the Pennsylvania Mine. Take a right at 39.6026 -105.8131 where you will see a directional sign. Follow that road across a bridge over Peru Creek to a fork in the road. Turn right to get to the mill site or left to go up the hill to the main mine portal and the top of the ore tram. To get to the mine up there you'll have to ignore the smaller roads cutting off of the main road and then turn left through an open gate, proceeding to a graded parking area at the top, next to the portal. Both the mill and the ore sorting house at the top of the tram line are still <u>impressive</u> buildings, each standing three stories tall. Highlights include the remaining aerial rail line that fed ore through the roof of the sorting house and also a couple of still-standing tram towers on the hillside between the sorting house and the mill. The aerial tram was used to move ore down to the mill. Neither building is safe to enter but people do sometimes explore them anyway. This is private property so do not remove anything at all, not even an ore sample. Treat the history and the private property with respect so we can all continue to enjoy this incredibly up-close experience. Before you continue uphill on the main road, feel free to explore the other, smaller roads on this hillside. You may find some of the upper workings of the Pennsylvania Mine or if you walk back west, the Callo Mine.

Continuing the tour upstream through the valley, past several mine dumps on the hillside, you reach the Argentine Pass Trailhead parking lot at 39.6088 -105.7999 on the left. Park here as there is a gate just uphill around the corner. Walk uphill, around the corner to see the Anderson Tun Mine with its impressive metal ore bin. Hike up above the ore bin to check out the rail lines leading back into the mine portal. From here, there are several options for hikes to additional mines further uphill. Following the old road uphill, the first turn off to the left leads first to the Peruvian Mine and then to the Whale Lode. If you continue straight to the next fork in the road, the left fork leads to the National Treasury Mine while the right fork leads to the Minikus, Olsen & Associates Claim - which is still an active mine (!!) so do not enter the active mining area. You will also see a couple other mines higher on the hills: on the right, well above the road is the North Snake River Mine and on the left the Paymaster, Peruvian and Shoe Basin Tunnel (this one has a trail running up to it). As I said before, this area is loaded with beautiful wildflowers and remote high mountain grandeur, so enjoy!

Upper Ten Mile Creek and McNulty Gulch

In 1861, a small group of miners came over the pass from Leadville into Upper Ten Mile Creek. They found good placer gold in McNulty Gulch and soon in several other gulches in the immediate area. Although this was in Summit County, there was no good route to the other towns in the county, so this discovery was fairly unknown to the gold miners in the Breckenridge area. In modern times, the large Climax Molybdenum mine has literally removed McNulty Gulch from existence with its open pit mining but there is still gold washed downstream for us to get. The original miners found the gold to be fine-grained and flakey which made it difficult to capture with primitive equipment. In many cases they ran the dirt several times and knew they were still losing a significant amount of the gold downstream. We have a chance to see some of their history in Mayflower Gulch, and to catch some of the gold they couldn't, by digging in upper Ten Mile Creek.

Mayflower Gulch Driving/Hiking Tour

Mayflower Gulch is up CO-91 south of Copper Mountain. You'll see the large, paved parking lot on the left at 39.4307, -106.1660, which is sadly lacking in signage. Park here unless you have a serious 4WD in which case you can drive up the first couple miles of trail to park near the old mining cabins. Otherwise, park and hike up the old mining road through the woods. Be sure to take note of the large ore bin on your right along the trail and a slowly collapsing miners' cabin on the left, nearby. Further up, as the trail reaches tree line, check out the old mining cabins, one even still has a roof. This area was the site of a concentrating mill, and you will also see some deposits of cast aside material. Most of the tailings here were cleaned up in the 1980s by a company that tried to reopen the hard rock gold mine further uphill. Sadly, gold prices dropped hard in the late 1980s and they gave up, eventually selling out to Summit County Open Space who now owns most of this land.

Hiking further uphill (and to the left a bit, not to the right) will take you through lots of summer wildflowers to a view, or even a visit, of the downhill end of a tram line which was historically used to move ore down from the mine portal FAR up on the hill. Can you see the portal way up there? Maybe following the cable up through the air to the cliff face will help in finding the old portal. Isn't it amazing that the cable is still in place running from the portal down to the tram base building? This mine is situated at over 12,000 feet in elevation!

Hiking further up the mining road will take you up a couple of switchbacks to the old dynamite storage (set into the mountainside) and then to the modern mine portal. Take a look inside to see just a little of the underground mine.

Prospecting Opportunities

When I wrote the *Finding Gold in Colorado: Prospector's Edition*, I was frustrated by the fact that both Breckenridge and Summit County ban all panning in their open space areas. While I still find that frustrating, I have now found enough other places that I feel pretty good about inviting people to come explore my home county with a gold pan. This book has almost three times as many dig sites in Chapter E as *Finding Gold in Colorado: Prospector's Edition* does!

Site Number: E-05A, B, C
Site Name: Ten Mile Creek

Ten Mile Creek runs down from the Climax Mine area to Copper Mountain Ski Area and then through Ten Mile Canyon to Frisco where it meets Lake Dillon. Along the way it passes the Mayflower Gulch area mentioned above. Some of the creek is off limits due to being on Climax Mine lands, within the no-prospecting zone around Dillon Reservoir, or on Summit County Open Space, but other parts are accessible due to being on either CDOT property or unclaimable USFS lands. Where the creek gets close to I-70/US-6 it is unclaimable (Public Land Order 3806, on 8/30/1965 withdrew all lands within 300' of US-6 from claimable status and this section of I-70 is also US-6), as is the creek near the ski area where the land was designated for recreational use only (PLO 3137, 7/30/1963). This combination gives us access to several patches of ground. As usual the sites listed here are from upstream to downstream. I should note that while I've found gold here, it has been quite sparse and small once you are alongside I-70. I hope your luck is better!

Local Hints & Cautions:
* Water levels in 10 Mile Creek vary seasonally making the creek essentially inaccessible during the late spring runoff but just great the rest of the season before and after that.

Gold Finding Tips:
* The gold here can be tiny and sparse. I generally found 1-2 specks per pan at best along I-70. Of course, your results may vary - I hope you find more and <u>bigger</u> stuff than I did!
* The gold here is usually but not always in the regular spots. Start by looking for bigger rocks on, or just downstream of, an inside bend. I found that most times, digging deeper led to less gold instead of more but your results may vary. If you run a sluice, sample pan regularly to confirm that you are still on the gold.

Getting There: This area runs along CO-91 from the lower edge of the Climax Mine property down to Copper Mountain's

Wheeler Junction. Then the creek runs along I-70 downstream to Frisco. Most of the creek has a recreational path along it which helps us with access. The sites are listed in my customary order from upstream to downstream but if you get off I-70 at exit 195 (CO-91) Site E-06B is close to the exit and then E-06A is a bit uphill along the first few miles of CO-91 heading southerly toward the Climax Mine up at the pass.

Locale: Summit County between Frisco and Freemont Pass (continuing over the pass, you'd find Leadville, Chapter G)
Land Type: moderate sized creek running through a narrow mountain valley and canyon
Land Manager: USFS, CDOT

Boundaries:
E-05A: From 39.4572, -106.1444 (just downstream of the Climax property) downstream, stop here or wherever there is an appropriate pull-off and give the creek a try. All of this section down to the next section described is unclaimable, but some of it is tricky to reach from the road. There are a couple of pull-offs marked "no parking", please respect that. Overall, using the rec path might be wiser and safer access for many. Obviously, while digging, don't do anything to increase risk of erosion to the rec path or the road.

E-05B: Access from Copper Mountain ski area "Far East" lot at 39.4973, 106.1373 with legal digging upstream into the prior site and downstream to 39.5041, -106.1406 where private property starts by the gas station.

E-05C: My experience is that downstream of the gas station property, 10 Mile Creek isn't worth working. I think this is because the beavers have dammed up the creek for thousands of years trapping the gold and building up lighter sediment. The creek has also been moved around somewhat for the freeway, US-6 before the freeway, and the railroad before that! Feel free to try the creek as it parallels I-70, where I have found a little color here and there, but you've been warned! NOTE: If you do explore that area, avoid the creek where it is close to the rec path (where you are unwelcome due to it being

Summit County property) and stay upstream of 39.5587, -106.1293. Downstream of that point, prospecting is forbidden by federal law (from 1961) to protect Dillon Reservoir.

Key Regulations:
• No gas-powered equipment.

Nearby Attractions & Accommodations:
Check out all the summertime fun at Center Village in Copper Mountain Resort. Toilets available there or at the gas station.

Prospecting the Upper Blue River

As I mentioned in the first book, Summit County and the town of Breckenridge bought up much of the old placer workings and made them into open space parks with no prospecting allowed. There is however one spot you can dip a pan.

Site Number: E-06
Site Name: CO-9 Bridge at Gold Hill

As described above, this is the only spot on the Blue River above Dillon Reservoir that is open to prospecting. It is very small but the access is easy and there's shade since some of the digging is under the state highway bridge.

Local Hints & Cautions:
• Avoid this site during the high water of spring runoff.

Gold Finding Tips:
• It's fairly easy to find some fine gold here with a pan. Sluicing should work well too but the gold is quite fine.

Getting There: From I-70 Exit 203 head south through Frisco on CO-9 to the parking just off CO-9 at Gold Hill Trailhead (39.5415, -106.0422). Then walk the paved rec path south a few dozen yards to the river and head east toward the highway bridge. Another option is to park just a little further south on the other side of the highway. Continue south on CO-9 for less than1/4 mile to the left turn onto Revett Drive (Sidebar: this road is named for a famous innovator of gold dredges who made a fortune dredging gold from the river here in the first half or the 20th century.). Follow Revett for a few hundred yards to a dirt lot at 39.5390, -106.0408 to park. Then just walk west across the open field to the dig area.

Boundaries: The stretch of river open to our access is just 165 feet long! The limits are upstream at 39.5394, -106.0422 to downstream at 39.5395, -106.0416 which is about 35 feet each way from the edges of the bridge.

Locale: Blue River between Frisco and Breckenridge
Land Type: modest sized river
Land Manager: CDOT

Key Regulations:
• No electric or gas-powered equipment is allowed here. Pans and sluices only.

Nearby Attractions & Accommodations:
There are lots of things to see and do in Breckenridge and Frisco. Camping is available at USFS campgrounds around Lake Dillon.

Prospecting the Lower Blue River
The Blue River flows through northern Summit County from Silverthorne to the north to fill Green Mountain Reservoir. It then continues north to meet the Colorado River in western Grand County on the edge of the town of Kremmling.

Site Number: E-07A, B
Site Name: Green Mountain Reservoir Inlet

The inlet of the reservoir is the point where the fast-moving river meets the still water of the reservoir. Where exactly that happens varies through the year as the reservoir level moves up and down. This reservoir moves up and down more than most because it is a power generation reservoir. The power plant is obliged to produce about $30,000 worth of electricity every day regardless of the current level of the water so sometimes it gets drained down a surprising amount. This matters to prospectors because when the water is low, we can access the reservoir/riverbed in areas that are sometimes the magic point where fast water comes to a stop and drops its load of gold (and cobbles as you will see if you visit when the water is low). Note: this site is downstream of site E-04 in the first guidebook.

Local Hints & Cautions:
- Remember that even when the reservoir is low, the river still has a lot of force. Be safe.
- This can be a good place to dig in the spring when river levels all over the state area are high. This is because the reservoir is often low in the spring after a long winter of power generation. On the other hand, when the reservoir is full in summer, this is a frustrating place to visit with very little access to paydirt. Come back in late fall when levels are lower so there is some reservoir bottom accessible.
- The banks are quite steep here so be careful!

Gold Finding Tips:
- Sample around thinking about where the gold would accumulate when this area is a river and where it would drop when the river is hitting the edge of the still reservoir at the area where you are standing.

Getting There:
E-07A: Grandview Cemetery: Follow CO-9 north from Silverthorne (I-70 Exit 205) to Heeney Rd. at 39.8240, -106.2155 which is about 16.6 miles and 20 minutes. Then follow Heeney Rd ½ mile to the parking on the right next to the old cemetery at 39.8293, -106.2198.

E-07B: The Upper West Side: Turn off of Heeney Rd. to the right at 39.8397, -106.2335 (down the road just a bit from the prior stop) and drive the dirt path into the parking lot. You can park wherever works for you as the whole USFS parking area is pretty informal.

Boundaries:

E-07A: As you work your way upstream, <u>stay</u> <u>away</u> from the east bank because that is the Blue River state wildlife area. Dig the WEST side of the river, to the center if you like, not more. Also avoid disturbing any anglers. The upstream boundary is the Bridge over the river, the downstream boundary is at 39.8309, -106.2197 after which there is private property for a stretch.

E-07B: Access here is from 39.8385, -106.2306 to as far downstream as you care to go within the bounds of the reservoir high water mark from bank to bank.

Locale: Green Mountain Reservoir near Heeney, between Silverthorne and Kremmling
Land Type: river running into a large reservoir
Land Manager: US Bureau of Reclamation

Key Regulations:
• Do not introduce new material into the reservoir. Only process material already in the bed of the reservoir.

Nearby Attractions & Accommodations:
Several camping options here and good fishing. Head south to Silverthorne for restaurants and craft breweries.

Site Number: E-08
Site Name: Prairie Point CG

In addition to being a campground, Prairie Point has a large day-use parking area (fee to use) in the perfect spot for us prospectors. There are also toilets.

Local Hints & Cautions:
• See points included in the prior site.

Gold Finding Tips:
- See prior site.

Getting There: There are two ways to play this. Either pay the fee and go to the day use parking area at 39.8419, -106.2327, or save a few bucks by walking from the free parking provided by the Bureau of Reclamation right off the highway just north of the turn off into the campground at 39.8433, -106.2315; obviously if you walk in, you shouldn't use the campground facilities such as the toilets. Either way you are on CO-9 until you pull off at the campground. This site involves driving past the turn-off to the prior site by the way.

Boundaries: Upstream access starts at 39.8404, -106.2306 where you will find a fence. Downstream you can go as far as you like. Additionally, down IN the reservoir this site goes upstream further to the point where it meets E-07B! As mentioned there, stay away from the east bank if you go upstream of the GPS coordinates above.

Locale: Green Mountain Reservoir near Heeney, between Silverthorne and Kremmling
Land Type: river running into a large reservoir
Land Manager: US Bureau of Reclamation, in collaboration with the USFS who runs the campground

Key Regulations:
- See prior site for details.

Nearby Attractions & Accommodations:
See prior site.

Site Number: E-09
Site Name: McDonalds Flats CG

This campground offers a gentler slope down to the water than the prior couple of sites. However, being further downstream, it offers access to the bottom of the reservoir less often.

Local Hints & Cautions:
- See points included in the prior site.

Gold Finding Tips:
- See prior site.

Getting There: Take the Heeney Rd. cutoff from CO-9 as mentioned in Site E-07 and continue further north to the turn off on the right at 39.8459, -106.2361 which is about 1.5 miles down Heeney Rd. from CO-9.

Boundaries: None relevant.

Locale: Green Mountain Reservoir near Heeney, between Silverthorne and Kremmling
Land Type: river running into a large reservoir
Land Manager: US Bureau of Reclamation, in collaboration with the USFS who runs the campground

Key Regulations:
- See prior site for details.

Nearby Attractions & Accommodations:
See prior site.

Site Number: E-10
Site Name: Green Mountain Power Plant day use area

An interesting little spot to get to since the drive takes you past the hydropower plant and some of its older, unused buildings. This site is just below the outflow of the dam, really the outflow of the hydropower plant.

Local Hints & Cautions:
- See points included in the prior site.
- This site is used by rafters so park in a way that will be out of their way if they show up with a truck and trailer.
- This spot is used by anglers, give them a friendly wave but do not speak loudly or approach if they are fishing.

- Be very careful getting up and down the hill from the parking to the river. There is usually a rope to hold onto for part of the journey, this is a good idea! This hill also means you will want to travel light or use some sort of sled to slide your gear up and down via a long rope or winch.

Gold Finding Tips:
- Water levels are fairly steady here through much of the year, but they can vary depending on the power plant output. Think about where that puts the gold.
- I found it easy to spot color in a test pan.

Getting There: From I-70 Exit 205, it is about 27 miles (34 minutes) to this site. As above, follow CO-9 through Silverthorne but this time head to the north end of Heeney Rd. at 39.8975, -106.3098 which is about 7.8 miles north of the prior turn off to Heeney Rd. From here, head west across the downstream end of the reservoir and across the dam (2.3 miles) then turn right onto CR-1812 past the empty guard shack for ½ mile to the day use area. On Google maps this spot is called Blue River launch below Green Mountain Reservoir.

Boundaries: Upstream below the dam where there is a fence to downstream several miles, far further than anyone is likely to walk, at 39.9120, -106.3411. This is theoretically about 3 miles of river but unless you are prepared to float it, only the upstream bit is realistic.

Locale: Downstream of the Green Mountain Reservoir near Heeney, between Silverthorne and Kremmling
Land Type: river running into a large reservoir
Land Manager: US Bureau of Reclamation and USFS

Key Regulations:
- Follow the rules posted on site.

Nearby Attractions & Accommodations:
See prior site.

Site Number: E-11A-C
Site Name: Blue River/Colorado River Confluence

The Blue River flows north out of Summit County into Grand County just southwest of the little town of Kremmling where it meets the Colorado River. The BLM recreation area at the confluence is only lightly used by anglers. It is more heavily used by rafters during periods of high water. When the water is low, there's almost no one there...except us!

Local Hints & Cautions:
- Water levels in the lower Blue River are driven by a combination of spring snow melt and water   releases for power generation at Green Mountain Reservoir. This latter source means water levels can vary unexpectedly.
- Take the stairs next to the boat ramp down to the river.

Gold Finding Tips:
- The gold here tends to be quite small, be prepared to catch the little stuff. All types of prospecting equipment are allowed here but places to run a sluice are rare/nonexistent and the gold is too fine for most dredges. A Gold Cube or a good production pan are the best choices in my experience.
- Focus on deposits of sandy gravel near shore in the riverbed. Bigger cobbles are a good sign. Avoid the clay, sand, and silt areas, they are barren.
- Densely packed material with a bit of muddy clay content mixed into the gravel and cobble seemed to pay the best (as is often the case!).
- Stick to the Blue River, the Colorado River access here is just mud and sand, not worth even a single test pan.

Getting There: From the intersection of CO-9 and Trough Road at 40.0364, -106.3742, turn west onto Trough Road, then right again following the BLM signage onto the northbound access road at 40.0309, -106.3868, then head up this dirt road.

Locale: Rural Grand County SW of Kremmling
Land Type: medium sized river running through a wide valley

Land Manager: BLM

Boundaries: There are three separate chunks of public access land along this road. Please respect the private ranch lands in between the BLM properties. In all cases here, access is to the near shore and river, but not to the far shore as that area is private ranchland.

NOTE: As of early 2025, access here will change in the near future for sites E-11A and 11B so watch for signage and fencing changes. A land swap has been approved between a local landowner, the BLM and the USFS along this section of river.

E-11A: 40.0324, -106.3860 to 40.0339, -106.3862 (Be sure to avoid blocking the road if you stop here to dig, ranchers use the road too.)

E-11B: 40.0360, -106.3881 to 40.0375, -106.3908 (with a nice pull off at 40.0379, -106.3901)

E-11C: The developed recreation area with parking and bathrooms at the end of the road. Prospecting area is from 40.0413, -106.3937 downstream to the confluence with the Colorado River. Parking for about six cars and bathrooms at 40.0429, -106.3954.

Key Regulations:
- Avoid blocking the boat ramp at the last site or leaving holes there while prospecting.
- All prospecting equipment is allowed here.

Nearby Attractions & Accommodations:
- Nearby Kremmling, a classic ranching town, has a grocery, a craft brewery/distillery (try the spiced whiskey!), a coffee shop and more. Blue Valley Spirits and Big Shooter's Coffee are always favorites for me but this tiny hamlet is the biggest town for 45 miles in any direction so there's just about one of everything here if you look around.
- If you enjoyed exploring this site, check out Trough Rd. in Chapter S for more sites and several campgrounds as well!

CHAPTER F: FAIRPLAY & PARK COUNTY

Alma and the Alma Placer

Across the river from the town of Alma is the Alma Placer, where it is easily visible to visitors. This site was first mined in the early years of the Colorado Gold Rush, with hydraulic mining starting here in 1869 after years of smaller scale operations. The first hydraulic-fed sluices were ¼ mile long! The hydraulic operation shut down in the early years of the 20th century but during the Great Depression as many as 300 men worked this area informally with hand tools and sluices.

The Alma Placer is at the meeting point of two ancient glaciers, one coming down from Buckskin Gulch and one that followed the current path of the South Platte River. Both glaciers carried gold into this area from 10,000 to 30,000 years ago. The dynamic nature of the glacial action means the gold deposits on the hillside are extremely erratic. The mine continues to operate in the modern era with a focus on gold but since they only get about one gram of gold per cubic yard of material, they are also selling all the sand, gravel, and cobble for other uses. As of 2021, they were processing 600 tons of material per hour and recycling all of their water resulting in zero discharge. The gold here is reported to be 82-84% pure with the rest mostly silver and copper as usual. Most of the gold is said to be 30-40 mesh. There are no public tours but you can see quite a bit of the operation from the highway.

Site Number: F-09
Site Name: Pennsylvania Creek Open Space

A very quirky spot; this 200-yard-long patch of Pennsylvania Creek was deeded back to the county when the neighborhood around it was created. I find this area fascinating because of the history of large nuggets on Pennsylvania Mountain, just to the west of here.

Local Hints & Cautions:
• You are in a neighborhood, basically in people's backyards so be a perfect guest. All it would take to lose this spot is a few homeowners complaining to the county.
• Access is from the neighborhood so be considerate about where you park.

Gold Finding Tips:
• I haven't dug this site myself (yet).
• Since this area is downstream of Pennsylvania Mountain, it would be wise to check your classifier for nuggets. It's unlikely but you never know!

Getting There: From CO-9 about halfway between Fairplay and Alma, turn west onto CR-1 at 39.2484, -106.0384. Follow that westerly to Pine Ridge Road. Take a right turn there and drive past 2 houses to the access path at 39.2491, -106.0596. This is between the second and third house on your right as you drive up Pine Ridge Road. From there, walk northeast between 68 Pine Ridge Rd. and 126 Pine Ridge Rd. to the creek.

Locale: Pennsylvania Creek between Alma and Fairplay
Land Type: modest creek running through a flat area
Land Manager: Park County

Boundaries: The upstream edge of is just west of the access path at 39.2500, -106.0592 with prospecting downstream to the southeast to 39.2493, -106.0573 with a caution that there is private property both upstream and downstream.

Key Regulations:
- This is a county open space, so no gas-powered equipment is allowed.
- NO digging into the slope of the creek banks or where rock has been added to stabilize the banks and to prevent erosion.
- Only dig the creek bed itself and leave things as you found them: knock your tailings back in your hole.

Nearby Attractions & Accommodations:
The towns of Alma and Fairplay have lots to offer. Start with South Park City if you haven't been to see this excellent living history museum. You drove past the next dig site on your way here so maybe you will have time to explore it too.

Site Number: F-10
Site Name: Platte River Ranch Estates Open Space

Like Pennsylvania Creek Open Space just west of here, this unbuildable land was deeded back to the county when the neighborhood around it was created. This open space property includes a nice stretch of the South Platte River: about 2/3 mile of riverbed.

Local Hints & Cautions:
- You are in a neighborhood, basically in people's backyards, so be a perfect guest. All it would take to lose this spot is a few homeowners complaining to the county.
- Some of the access is from the neighborhood street so be considerate about where you park.

Gold Finding Tips:
- I haven't dug this site myself (yet).
- There are some nice little bends in the river here to check.

Getting There: From CO-9 about halfway between Fairplay and Alma, turn west onto CR-1 at 39.2484, -106.0384. Follow that westerly 0.2 miles to Cottage Grove Road. Take a right turn there and drive along this road to any likely parking spot.

Locale: South Platte River in western Park County
Land Type: medium sized river running through a flat area
Land Manager: Park County

Boundaries: Basically, the whole stretch of river along Cottage Grove Road is open access. The upstream edge of is at 39.2592, -106.0482 with prospecting downstream to the south to 39.2504, -106.0469 where the river meets CR-1. Remember the caution that there is private property both upstream (someone's backyard) and downstream (a ranch).

Key Regulations:
- This is a county open space, so no gas-powered equipment is allowed.
- NO digging into the slope of the creek banks or where rock has been added to stabilize the banks and to prevent erosion.
- Only dig the creek bed itself and leave things as you found them: knock your tailings back in your hole.

Nearby Attractions & Accommodations:
The previous dig site is very close by.

Site Number: F-11
Site Name: Sacramento Creek west of Fairplay

This mile long stretch of Sacramento Creek is a bit downstream of where the Hoffmans mined this creek on the Gold Rush TV show. This section of the creek was made unclaimable quite a long time ago via Public Land Order 3149.

Local Hints & Cautions:

- Stay away from the private property upstream and downstream.

Gold Finding Tips:
- I haven't dug this site myself (yet).
- There are lots of little bends in the creek along this stretch.
- Avoid the beaver ponds as mentioned below. Instead, dig where the creek is flowing and you can see gravel and cobble.

Getting There: From Fairplay, head north on CO-9 for about two miles or so to Sacramento Creek Road at 39.2386, -106.0301 then follow Sacramento Creek Road west to the prospecting area. You'll know you are getting close when the pavement ends! Since you are driving upstream as you approach the site, you will first arrive at site C (listed below).

Locale: Sacramento Creek just upstream (west) of Fairplay
Land Type: modest sized creek
Land Manager: USFS

Boundaries: The creek runs along the south side of Sacramento Creek Road, flowing from west to east. The upstream edge of the one-mile-long prospecting area is at 39.2240, -106.0690 with prospecting downstream to the east to 39.2254, -106.0503. Since the creek runs from east to west, you can just focus on the longitude numbers to keep track of where you are here. Remember the caution that there is private property both upstream and downstream. On the upstream end, do NOT use Gold Pan Lane to access the prospecting area. The land on both sides of the lane is private.

There are lots of places to park along the side of Sacramento Creek Road without blocking the roadway. It's a short walk from any point along the road to the creek. However, it is worth highlighting several informal roads that lead close to the creek from upstream to downstream.

F-11A: 39.2251, -106.0646 This road heads south then splits with both forks leading very close to the river.

F-11B: 39.2254, -106.0634 This road drops southerly to the river then sweeps west to meet the road described just above. The stretch of creek between sites B and C is very boggy and filled with beaver ponds so it is unlikely to be productive unless you are willing to dig past a lot of sedimentation and silt. Stay away from any active beaver ponds.

F-11C: 39.2256, -106.0560 This road only drops south a bit, then runs parallel to Sacramento Creek Road, heading east. It runs almost all the way to the eastern boundary. Past the end of this road there is another beaver pond to avoid.

Key Regulations:
• Stay away from active beaver habitat, they are protected.

Nearby Attractions & Accommodations:
See notes on previous dig sites. Also, this site is very close to South Park City living history museum which is definitely worth a visit.

Site Number: F-12
Site Name: South Platte above Fairplay

This odd little piece of county land is between the town of Fairplay and the site where the Hoffman Crew mined on the Gold Rush TV show some years ago.

Local Hints & Cautions:
• You are very visible here so please be friendly to the tourists and locals who stop to ask questions.

Gold Finding Tips:
• I haven't dug this site myself (yet).
• There is a nice little bend in the river on the upstream part of this property.

Getting There: From CO-9 & US-285 go south over the river on US-285 to Platte Drive. Turn right onto Platte Drive and follow it about a mile to the pull off on the right at 39.2236, -106.0083. Walk about ¼ mile northwest to the river.

Locale: South Platte River just upstream of Fairplay
Land Type: medium sized river
Land Manager: Park County

Boundaries: The river runs along the north edge of this property. Stay on the south side of the river and in the riverbed since ownership of the north shore is private. The upstream edge of is at 39.2276, -106.0109 with prospecting downstream to the southeast to 39.2265, -106.0086. Remember the caution that there is private property both upstream (an active mine!) and downstream (town property).

Key Regulations:
• This is county open space so no gas-powered equipment is allowed.
• NO digging into the slope of the creek banks or where rock has been added to stabilize the banks and to prevent erosion.
• Only dig the creek bed itself and leave things as you found them: knock your tailings back in your hole.

Nearby Attractions & Accommodations:
See notes on previous dig sites. Also, this site is very close to South Park City living history museum which is definitely worth a visit. Site F-01, F-02 and F-03 from the original *Prospector's Edition* are just downstream of here to the south.

Site Number: F-13
Site Name: South Platte at Middle Fork Vista

This odd piece of county land is between a ranch and a fairly newly developed neighborhood. When the neighborhood was platted, the unbuildable land next to the river was apparently deeded back to the county.

Local Hints & Cautions:
- Be careful going down the steep hill to the river.

Gold Finding Tips:
- I haven't dug this site myself (yet).
- The river has many bends in it here so there is plenty to explore along the 0.8 miles of accessible river.

Getting There: From the south end of Fairplay at CO-9 & US-285, go south over the river on US-285 for a mile to the left turn onto CO-9. Follow that 7.4 miles to the left turn onto Redhill Rd (CR-7) for less than ¼ mile and then take the hard left to stay on Redhill Rd. Follow that 1.1 miles to the slight left onto Trout Creek Ln., then 0.9 miles on Trout Creek to the slight left onto Middle Fork Vista. Take that about 1.4 miles and park at 39.1657, -105.9374. The empty lot on the west side of the road is the public access to the open space area. Be sure to park fully off of the neighborhood road. Walk north on the road to 39.1676, -105.9382 and follow the electrical lines west into the open space. Walk northwest to the downstream end of the river access at 39.1713, -105.9452 where you will see a barren area where the river used to run.

Locale: South Platte River just upstream of Fairplay
Land Type: medium sized river
Land Manager: Park County

Boundaries: The river runs along the west edge of the open space property. The upstream edge of is at 39.1826, -105.9461 with prospecting downstream to the starting point of your hike along the river at 39.1713, -105.9452 as mentioned above. Remember the caution that there is private ranch land both upstream and downstream.

Key Regulations:
- This is a county open space so no gas-powered equipment.
- Do not access from CO-9, this is for homeowners only.

Nearby Attractions & Accommodations:
See notes on previous dig sites. Also, this site is very close to

South Park City living history museum which is definitely worth a visit.

Site Number: F-14
Site Name: Horseshoe Campground

This campground is on the south side of Four Mile Creek, a little bit southwest of Fairplay. With historic mining sites upstream, you can be sure there is a chance for gold here.

Local Hints & Cautions:
- Some of this area is quite boggy, especially in spring. Watch your step or come later in the season when things dry out.
- Fishing is popular here. Give any active anglers a wide berth to avoid scaring the fish they are pursuing.

Gold Finding Tips:
- Look for surface gravels and cobbles. Dig down into the hardpack material.

Getting There: From Fairplay, head south on US-285 for 1.4 miles, turning west onto CR-18 (Four Mile Road) and following that for 8 miles. The campground is at 39.2012, -106.0814, on the south side of the highway. Look for a sign several hundred yards before the entrance. Ask the campground host about day-use parking and fees if you don't plan to camp here...or just park off the road and walk in from CR-18.

Locale: Four Mile Creek in western Park County
Land Type: medium creek running through a flat area
Land Manager: USFS

Boundaries: The upstream edge of is at 39.2018, -106.0846 with prospecting downstream to the east to 39.1996, -106.0809. Don't wander off of the public access area, there are active placer claims both upstream and downstream.

Key Regulations:
- There is a fee for day-use or camping here.

- This is a campground, so no power equipment is allowed.
- NO digging into the slope of the creek banks or where rock has been added to stabilize the banks and to prevent erosion (such as near the roads).

Nearby Attractions & Accommodations:
Camp on site, reservations recommended: recreation.gov or 877-444-6777. There are 19 sites with parking spurs 25-40 feet long. Vault toilets and hand pumped water available in a shared area.

Alternatively, camp a little further west along CR-18 at Four Mile Campground on the north side of the road. Camping here is "first come, first served" with 14 campsites. The campground is open from May to early October. Vault toilets and hand pumped water are provided at a shared facility. Maximum trailer/RV length of 22 feet.

Site Number: F-15A-D
Site Name: Geneva Creek

There are several portions of Geneva Creek, in northeast Park County that are unclaimable and open to us as casual prospectors. This is a beautiful area with lots to explore, places to stay, and other fun outdoor activity.

Local Hints & Cautions:
- Fishing and hiking are popular here. Give any active anglers a wide berth to avoid scaring the fish they are pursuing. Be friendly with the hikers, they may be the next gold prospectors!
- Replace your tailings in your hole as much as practical to preserve the pristine appearance of the creek for other uses.
- This road is closed for the winters – seasonal access only.

Gold Finding Tips:
- I haven't explored this area myself; the reports of gold are from a friend. I expect this to be a challenging place to find

gold so be prepared to be tenacious and look for the smaller specks.

Getting There:
From Fairplay, head north toward Denver on US-285 for 27.8 miles, turning north onto Geneva Creek Road (CR-62), aka Guanella Pass Road, at 39.4597, -105.6631, and following that for 9.1 miles to the dig site which is furthest upstream (see site A below). Along the way you may see signs indicating Duck Creek Road, FR-119, or CR-1038 – these are all the same road for our purposes.

Locale: Geneva Creek in northeastern Park County
Land Type: medium sized creek
Land Manager: USFS

Boundaries:
Site F-15A Kirby Gulch: Park near the upstream edge if this first area at 39.5532, -105.7547 and walk downhill, west to Geneva Creek with prospecting from 39.5530, -105.7573 downstream to the southeast to 39.5433, -105.7469 and also in Kirby Gulch from 39.5465, -105.7573 to the confluence with Geneva Creek. There are additional pull off areas along the road in this mile long stretch of accessible creek so feel free to park in other spots along this stretch for a shorter walk to various parts of the creek here.

Site F-15B Geneva Park CG & Burning Bear CG: This area is a natural choice if you are staying in either of the campgrounds. Geneva Park CG is near the north end of this site, on the west side of the road, at 39.5311, -105.7332 and Burning Bear CG is at the south end, on the east side, at 39.5131, -105.7127. Otherwise perhaps you'll want to park in a wide spot along the road or at the Duck Creek Picnic Area a little downstream of Geneva Park. This stretch of Geneva Creek prospecting includes a little over 2 miles of waterway from just upstream of the campground at 39.5316, -105.7347 downstream to the southeast to 39.5125, -105.7208 which is due west of Burning Bear USFS CG. You can also prospect in Bruno Gulch from 39.5231, -105.7356 to the confluence with Geneva Creek and in

Burning Bear Creek from 39.5120, -105.7387 to its confluence with Geneva Creek. Both of these creeks run from the hills west of Geneva Creek, eastward to meet it.

> *NOTE: Finding Geneva Park USFS CG is tricky- Look for it on the right as you head downhill. The signage for this CG only faces the road for those driving uphill, so keep a sharp eye out otherwise. If you get to the Duck Creek Picnic Area going downhill, you've gone past the campground.*

Site F-15C Geneva Creek: Geneva Creek prospecting from 39.5062, -105.7088 downstream for ¾ mile to the southeast to 39.4977, -105.6987 which is just upstream of Tumbling River Ranch. This all-inclusive dude ranch is known for its horse programs but also much more - www.tumblingriver.com for more. There isn't any good parking at the upstream end of this section, park on the west side of the road at 39.5053, -105.7049 or see what you can find along the side of the road from there downhill to the ranch.

Site F-15D: Three Mile Creek Trailhead: Geneva   Creek prospecting from 39.4818, -105.6937, near the trailhead which offers a bridge across the creek and parking, downstream to the southeast to 39.4743, -105.6846, near Geneva Creek Picnic Area where there is also a bridge over the creek and some parking. Stay away from the bridges when prospecting. Parking between the formal parking areas may be challenging.

Key Regulations:
- In a few places the creek runs close to the road. There and elsewhere be sure to avoid any installed structures or erosion control rockwork.
- Only dig IN the creek bed and cobble bars, never into or on the banks.

Nearby Attractions & Accommodations:
Head east a little bit from here along US-285 to Grant and Bailey to check out the towns (the South Park Coney Island Boardwalk west of Bailey is an iconic Colorado stop), or west to Fairplay of course. On the way toward Fairplay, you'll pass by the turn off to the next dig site too.

Site Number: F-16
Site Name: Ute Creek Trailhead

This Trailhead is just off of Tarryall Road. A portion of the creek here is on unclaimable forest service land between two sections of private property, so we get our shot at it.

Local Hints & Cautions:
- Fishing and hiking are popular here. Give any active anglers a wide berth to avoid scaring the fish they are pursuing. Be friendly with the hikers, they may be the next gold prospectors!
- Do not dig near the pedestrian bridge over the creek.

Gold Finding Tips:
- There are several nice inside bends and cobble bars here to explore.

Getting There: From Fairplay, head north toward Denver on US-285 for 16 miles, turning south onto Tarryall Road (CR-77) at the little hamlet of Jefferson and following that for 20.6 miles. The trailhead parking is at 9814 Tarryall Road or 39.1980, -105.5537, on the northeast side of the road. There is parking for quite a few cars here, but it can get busy on weekends during hiking season.

Locale: Tarryall Creek in eastern Park County
Land Type: medium sized creek running through a flat area
Land Manager: USFS

Boundaries: The upstream edge of is just north of the parking loop at 39.1984, -105.5538 with prospecting downstream to the east to 39.1965, -105.5497. Don't wander off of the public access area, there are private ranches both upstream and downstream.

Key Regulations:
- NO digging within 30 feet of Tarryall Road. (This is only relevant at a couple of spots where the road and creek almost meet, downstream of the parking area.)
- When near the bridge over the creek only panning and sluicing are allowed. No power equipment within 50 feet of the bridge.

Nearby Attractions & Accommodations:
Try hiking Ute Creek Trail. There is camping 5.6 miles south on Tarryall Road at X-Rock USGS CG or just a bit further south at Spruce Grove CG (both of those campgrounds are prospecting sites in *Finding Gold in Colorado: Prospector's Edition*). To the north, there is camping at Tarryall Reservoir. Those camping areas require a valid hunting or fishing license or an SWA pass (available at local retailers for $9/person/day or about $50 per season) at
https://cpw.state.co.us/aboutus/Pages/Fees.aspx#SWAFees
or by phone 800-244-5613. At least there is no additional fee to camp if you have your State Wildlife Area passes. The next site is on a different waterway but the access is just a bit south of this site on Tarryall Road.

Site Number: F-17
Site Name: Happy Meadows Rec Area

The Happy Meadows area includes a USFS Campground and over two miles of the South Platte River. He encourages you to look for amazonite (light green milky colored semi-precious mineral) and petrified wood here. You will see them in the river cobbles as you dig. Public Land Order (PLO) 1793 made a very large area here unclaimable so there is lots to explore.

Local Hints & Cautions:
- Fishing is very popular here. Give any active anglers a wide berth to avoid scaring the fish they are pursuing. Visiting during the week when there are fewer people will reduce this issue.

- CR-112 follows along next to the river through this whole stretch with multiple pull-offs along the way, so access is very easy.

Gold Finding Tips:
- There are several nice inside bends and cobble bars here to explore. In fact, with over two MILES of river, there's just a lot of everything to explore!

Getting There: The town of Lake George in southern Park County, on US-24, head north on CR-77/Tarryall Road for about a mile to CR-112. Then travel east on CR-112 for about a mile to the campground.

Locale: South Platte River in southeastern Park County
Land Type: medium sized river running through a flat area
Land Manager: USFS

Boundaries: The south boundary is not far after you get off of CR-77 at 39.0083, -105.3644 with prospecting downstream to the north, well past the campground to 39.0261, -105.3512. Don't wander off of the public access area, there are private ranches both upstream and downstream. Then again, with over two miles to explore here, you won't likely be tempted to get that far.

Key Regulations:
- No gas-powered equipment within 30 feet of the centerline of the road.
- Fill all holes to protect the safety of the anglers.

Nearby Attractions & Accommodations:
Stay at the Happy Meadows CG, visit the little shops in Lake George or play on the lake. Drive east on US-24 to visit Florissant Fossil Beds National Monument with its fossilized Redwood tree stumps and other fossils.

The next sites downstream on the South Platte River are in back in Chapter A because they are in Jefferson and Douglas Counties which I categorize as part of metro Denver.

CHAPTER G: LEADVILLE AND THE UPPER ARKANSAS RIVER BASIN

The Arkansas River didn't create the valley it flows through. The valley is actually a geologic rift created when two tectonic plates were "welded" together during the formation of the North American continent. The river was pushed into the eastern edge of the rift valley by large deposits of glacial debris which now form the gold bearing lower slopes of the west side of the valley.

During the ice ages, the river was dammed by glaciers near the modern Clear Creek Reservoir. This formed a 600-foot-deep lake north of the ice dam. The dam broke and reformed at least three times with the most recent being about 17,000 years ago. Each break led to a dramatic flooding event pushing gold downstream, up onto hillsides along the river and even onto the eastern plains. Driving near Buena Vista, the car sized rocks sitting on the landscape are a reminder of the giant floods but the gold in our pans is a better reminder!

Prospecting Opportunities
The following prospecting sites are a mix of smaller feeder creeks that eventually flow into the Arkansas River, as well as a couple more prospecting sites on the river itself. I left those sites out of the prior book because of some ambiguity around mining claims that had been filed there but those claims were invalid because the area is, indeed, unclaimable.

Site Number: G-05
Site Name: Elbert Creek Campground

Elbert Creek Campground is in a little, wooded valley between Mt. Massive and Mt. Elbert, two of Colorado's most famous 14,000+ foot mountains. The USFS warns that altitude sickness can occur at this elevation. The Continental Divide Trail, which is also the Colorado Trail in this area, crosses the road just west of the campground. Half Moon Creek flows eastward along the north edge of the campground.

Local Hints & Cautions:
- The campground is "first come/first served". Do not park at a campsite unless you pay for it.

Gold Finding Tips:
- Look for fine gold in the sticky paydirt.

Getting There: From Leadville, travel south approximately 3 miles on US Hwy 24. Turn RIGHT on Colorado 300 at the Leadville National Fish Hatchery sign. Continue about 3/4 of a mile and turn LEFT at the Halfmoon Campground sign. Follow this road and turn RIGHT at the Halfmoon sign at the end of the pavement. This dirt road will lead to the Halfmoon East/West CGs and Elbert Creek CG areas. This site is Elbert Creek Campground of course. (Note: This road is not maintained in the winter.)

Boundaries: Upstream access starts at the west edge of the campground at 39.1523, -106.4191 and the downstream limit is 39.1546, -106.4077. Since the creek and road run east-west, just keep an eye on the longitude to stay on legal ground. The road runs along next to the creek over this whole distance of about 3/4 mile. There may be wide shoulders to pull off along the way, I didn't take notes on that when I visited.

Locale: Between Mt Massive and Mt Elbert, south of Leadville
Land Type: modest sized creek, in a mountain valley
Land Manager: USFS

Key Regulations:
- No gas-powered equipment in the campground.

Nearby Attractions & Accommodations:
Camping, hiking, climbing 14ers like Mt. Massive and Mt. Elbert. Exploring Leadville, including the National Mining Hall of Fame and Museum. There is a lot to do around here!

Site Number: G-06
Site Name: Half Moon East Campground

Half Moon East Campground is just about a mile east of the prior site so you will drive past it on the way to Elbert Creek CG if you are headed there. Half Moon Creek flows northeastward along the far edge of the campground.

Local Hints & Cautions:
- During some seasons there are a lot of muddy, marshy areas along this creek. It is best to avoid those areas both for environmental reasons and because that isn't where the gold collects.
- Do not park at a campsite unless you pay for it.

Gold Finding Tips:
- Look for fine gold in the sticky paydirt.

Getting There: From Leadville, travel south approximately 3 miles on US Hwy 24. Turn RIGHT on Colorado 300 at the Leadville National Fish Hatchery sign. Continue about ¾ of a mile and turn LEFT at the Halfmoon Campground sign. Follow this road and turn RIGHT at the Halfmoon sign at the end of the pavement. This dirt road will lead to the Halfmoon East/West CGs and Elbert Creek CG areas. This site is Half Moon East Campground of course. (Note: This road is not maintained in the winter.)

Boundaries: Upstream access starts at the south edge of the campground at 39.1578, -106.3995 and the downstream limit is

39.1828, -106.3856. The road runs along next to the creek over this whole distance of about 2 miles. There may be wide shoulders to pull off along the way, I didn't take notes on that when I visited.

Locale: Between Mt. Massive and Mt. Elbert, south of Leadville
Land Type: modest sized creek, swampy in parts, in a mountain valley
Land Manager: USFS

Key Regulations:
• No gas-powered equipment in the campground. Permissible once north of 39.1623, -106.3952.

Nearby Attractions & Accommodations:
Camping, hiking, climbing 14ers like Mt. Massive and Mt. Elbert. Exploring Leadville, including the National Mining Hall of Fame and Museum. There is a lot to do around here!

Twin Lakes & Lake Creek

The little town of Twin Lakes is just up CO-82 at the upstream end of the Twin Lakes reservoir. It started out as a stage stop for the stage line between Leadville and Aspen when both towns were booming with miners and were making millionaires of some of the luckiest prospectors. Twin Lakes quickly developed into a tourist destination because of its beautiful location on the shores of the two largest glacial lakes in Colorado. There was a fancy hotel for the swanky folks on the south side of the lake...and a brothel and bars on the north side. Today the town (on the north side of the lake) has quite a few preserved buildings, a general store, and a restaurant/tavern/hotel which is still in operation. You'll also find a couple of shops to browse in or to grab a coffee. Twin Lakes is known as a speed trap and is famous for having a parked sheriff's car on the main road through town with a dummy in the driver's seat. There's even a bag of donuts on the dash! If you look closely, you'll see the license plates on the patrol car expired in 2004! There is also often a real patrol car in the area, so do watch your speed in town. If you want to see the old hotel and walk through a nicely restored, but unfurnished, summer home, you'll have to drive to the west edge of the dispersed camping area and then hike a couple miles west on the trail just above the lake shore. It's pretty walk and relatively flat so I recommend it. Just north of here is the mining boom town of Leadville with many more attractions (Details in the book *Finding Gold in Colorado*). Just to the south is the Cache Creek prospecting site and the Lost Canyon Road driving tour.

Further along the road is Independence Pass. Just over the pass is the ghost town of Independence and then, of course, Aspen. (see Chapter W)

While the little town of Twin Lakes and the Lake Creek drainage was not known for a boom of its own, many prospectors passed through here and of course some dipped a pan in the local waters. Although the gold was not generally rich enough to justify commercial mining, there is gold to be found at the public access areas.

Site Number: G-07
Site Name: Perry Park Campground

This campground is perched on the benches in the canyon of Lake Creek upstream a few miles from Twin Lakes.

Local Hints & Cautions:
- There is an old stagecoach road on the south side of the creek. Access it from the campground for easy hiking access downstream.
- The water here is extremely fast and high during spring runoff. It rises quite a bit after summer rains as well. During those periods, access to gold bearing paydirt in the creek is limited. Be warned: people drown in Lake Creek every spring, be careful!
- Some of this area is a very pretty canyon, worth visiting just for the beauty. Be careful as a prospector!

Gold Finding Tips:

- The gold here is mostly quite fine. In my sampling, everything was -50 to 200 mesh. A good pan had at least 6-8 colors in it during my testing.
- Inside bends and areas with larger rocks were better but sample around because my results varied quite a bit. Some spots that looked very good produced only a speck here and there, others were oddly better.
- Bedrock crevicing is fun here and can be productive. Be sure to properly dispose of any lead you find in those crevices.

Getting There: From Leadville, head south on US-24, turning west onto CO-82. The campground is at 39.067680 -106.409780, on the south side of the highway. Look for a sign several hundred yards before the entrance. Ask the campground host about day-use parking and fees if you don't plan to camp here. Of course, you can also look for a wide shoulder to park on where the road parallels the creek.

Locale: Twin Lakes in southern Lake County
Land Type: large creek running through a canyon
Land Manager: USFS

Boundaries: The upstream edge of this area is a private property boundary just upstream of the campground at 39.0676, -106.4117. The downstream edge of the prospecting access is a decent way below the campground at 39.0625, -106.4067. Since the creek runs west to east you can just watch your longitude as you explore.

Key Regulations:
- There is a fee for day-use or camping here.
- This is a campground, so no power equipment is allowed.
- NO digging into the slope of the creek banks or where rock has been added to stabilize the banks and to prevent erosion (such as around the bridge).

Nearby Attractions & Accommodations:
The little town of Twin Lakes is downstream of here. There's not much to it but it's worth checking out. Be sure to watch your speed through town. Sometimes the cop car has a dummy in it (LOL!) and sometimes there is a real officer with a speed gun!

Site Number: G-08
Site Name: Twin Lakes

Lake Creek was never heavily placer mined but the traces of gold in the creek did lead prospectors to several viable hard rock mine sites in the hills upstream. There is also gold deposited here by glacial action with unknown sources. Below the Twin Lakes dam, the creek flows through a dispersed camping area and a basically flat plain. The gentle flowing water here (even before there was a dam) means this is an accumulation zone for gold...but not a sluicing spot.

Local Hints & Cautions:
- The water here is relatively high during runoff and through mid-August as water is released for rafting and farmers downstream. However, even during those periods, access to gold bearing paydirt in the creek is decent.

Gold Finding Tips:
- The gold here is mostly quite fine. In my sampling, everything was -50 to 200 mesh. A good pan had at least 6-8 colors in it during my testing. Inside bends and areas with larger cobble seemed to pay the best but be sure to sample around because my results varied quite a bit.
- Bring your heavy panning concentrates home for final careful processing. This gold is small so you will find more hiding in the black sands if you process it carefully at home.

Getting There: From Leadville, head south on US-24, turning west onto CO-82. The turn into the area below the dam is only about a mile off of US-24. Turn left onto CR-25, drive a few hundred yards on the dirt road and you will see dispersed camping sites on the right side of the road. Many of these sites are very close to the creek and provide good access to the best panning areas.

Locale: Twin Lakes in southern Lake County
Land Type: large creek running through a flood plain
Land Manager: US Bureau of Reclamation

Boundaries: The upstream edge of this area includes various facilities related to the dam. Stay away from those improvements. The downstream edge of the prospecting access is at the road bridge.

Key Regulations:
- Respect signage and fences.
- Do not dig near the highway bridge or other improvements such as those related to the dam.
- NO digging into the slope of the creek where rock has been added to stabilize the banks and to prevent erosion.

Nearby Attractions & Accommodations: See below.
See comments on the prior site. There is lots of space here for dispersed camping. Kayaking on the lake is also an attraction.

Driving Tour: Lost Canyon Road
This is a four-wheel-drive sightseeing tour with no prospecting opportunities once you pass the BLM Cache Creek property turnoff. Lost Canyon Road (FR-398) starts in the tiny town of Granite on US-24 between Leadville and Buena Vista. It is just a couple miles south of CO-82 where Lake Creek and Twin Lakes are located. Lost Canyon Road heads west into the hills, first passing the Cache Creek prospecting site (described in detail on my website www.findinggoldincolorado.com and in my first book *Finding Gold in Colorado: Prospector's Edition*).

Getting There: From Leadville, head south on US-24, turning west onto Lost Canyon Road at the sign in Granite. From Buena Vista, head north on US-24 to Granite.

The Driving Tour:
The good quality, but narrow, dirt road climbs quickly onto a plateau and continues west to some high-power electrical line. Turning left here, paralleling the power line will bring you to the Cache Creek prospecting area in about a mile (rough road in spots). Continuing on Lost Canyon Road, FR-398, the road gets more challenging and really requires a high clearance vehicle at times. I recommend a 4WD truck if you want to do the tour.

The next section of the drive winds uphill through pretty pine and aspen forests, affording occasional views across the valley to the north or south.

Proceeding further you will start to see old mining buildings and vehicles. Please respect that all of this is private property: view only from the road. In fact, the mine is still operational during the summer months. The private mine includes all of the creek from this point upstream to the headwaters. Happily, it is easy to see the mining structures and operations from the road.

Further along, an experienced prospector's eye will see more evidence of mining in the narrow valley, from large placer tailing piles to areas of the bank and hillside which have been dug back to get to the hillside paydirt. Keep an eye out for a tall metal structure, this is the remains of a small bucket line, dry land dredging operation which dug into the hillside along the creek. A conveyor system (now long gone) carried paydirt back to the creek for processing. Notice the artificial cliffs in this area where the entire hillside has been removed to be processed for its gold.

Next several old log cabins are accessible along the road. Feel free to have a look but do not take anything, not even a stray nail, as this is all protected by law. More evidence of mining is easy to notice down in the creek.

The end of our drive is up above tree line. Looking along the creek to the west, the headwaters are visible with additional mining tailings piles and more modern mining buildings. The other highlight here is the grand views of Twin Lakes, and even Leadville's high valley from your perch atop the mountain. It is rare to be able to drive this high. Multiple ranges of mountains are visible including quite a few 14,000+ foot high peaks. The view is undoubtedly one of the grandest in the Rocky Mountains!

Locale: Lost Canyon is along the boundary between southern Lake County and northern Chaffee County

Land Type: 4WD road along a hillside and creek, culminating above tree line

Land Manager: USFS, with private land adjacent to the road

Key Regulations:
- Only drive on the existing road.
- Stay off of the private mining area, even where it is unmarked.

Nearby Attractions & Accommodations:
The little town of Twin Lakes is just up CO-82 and the beautiful glacial lakes of Twin Lakes as well. Dispersed camping is available on the high plateau at the beginning of this drive. Lots of more developed camping is available just uphill from Twin Lakes. This includes both dispersed camping and USFS campgrounds. There is also a small hotel in Twin Lakes.

Prospecting Opportunities

Options abound in this area including Cache Creek, sites along the Arkansas River and along Lake Creek (which feeds Twin Lakes and then continues into the Arkansas River. The Cache Creek site is described on my free website as are some of the Arkansas River sites. Other Arkansas River sites are in the book *Finding Gold in Colorado: Prospector's Edition* while the Lake Creek sites are earlier in this chapter. Just south of here, Clear Creek reservoir offers additional dispersed camping on its western shores, more prospecting and another sight-seeing drive. That one goes through several small ghost towns and is also detailed in *Finding Gold in Colorado: Prospector's Edition*.

The numbering on the two following sites is a continuation of the numbering of sites along the Arkansas River in the prior guidebook. The first two sites are just a bit upstream (northerly) from Buena Vista aka "BV" and is downstream of site G-01G in Volume One.

Site Number: G-01H
Site Name: Dude, where's my reservoir?

This area is unclaimable due to being withdrawn in 1894 for a planned reservoir, which doesn't exist. The site numbering is because this site belongs next in the G-01A thru G-01G sequence in the original edition of *Finding Gold in Colorado: Prospector's Edition*. This site was excluded from the original book because I hadn't confirmed the legal status of the land in time for publication – there was an active claim here at the time. The BLM eventually notified the claimant that this land is unclaimable, and revoked the claim, so here you go.

NOTE: you may still see the old claim markers, but you can confidently ignore them.

Local Hints & Cautions:
• Just the usual comments about the Arkansas River: be wary of changes in river levels which may be caused by releases from the Twin Lakes dam upstream or rainstorms.
• The river runs basically north to south through here so use latitude numbers to gauge when you are on public land.

Gold Finding Tips:
• The prospecting here is fairly typical for the upper Arkansas River. Fine gold collects in typical locations, as in any river, but can be spotty.

Getting There: Head north from Buena Vista or south from Leadville on US-24 to small gravel parking on the east side of the highway at 38.8901, -106.1691 which is just upstream of the valid public prospecting area. Alternatively, park anywhere you see fit along the east side of the highway (where you can be far enough off the road to be safe) from the gravel lot for 1/3 of a mile north. This 1/3 mile is all reacquired federal land on the east side of the road. Either way, do not block the gate or impede the ability of the local residents to get in and out of their property.

Walk to the north end of the little parking area and take the trail down to the water following the "river access" signage. This access is on public land, but the stairs are maintained by the residents next door, so please treat it with respect.

If you parked along the highway north of the gravel lot, feel free to cross public land directly from your vehicle to the river. The public lands extend along the highway from the gravel lot north to 38.8941, -106.1732 ...north of that there is private property between the highway and the river, but the river is still public for quite a way further north to 38.9018, -106.1727 for a total of 1.25 miles of river!

Locale: just north of Buena Vista in Chaffee County
Land Type: medium sized river in a mountain valley
Land Manager: BLM

Boundaries: The upstream edge of this area is a private property boundary at 38.8941, -106.1732. The downstream edge of the prospecting access is 38.8868, -106.1640 at a private property boundary.

Key Regulations:
• A permit is required to use power equipment of any sort on the Arkansas River here. The Arkansas Headwaters Recreation Area office in Salida provides the permits.

Nearby Attractions & Accommodations:
Buena Vista is just down the road. There are lots of RV parks and other travel amenities in this area. Eddyline Brewery in BV is a favorite for local craft beer.

Site Number: G-01I
Site Name: Mickey Mouse BLM Site

No, I don't know why the BLM calls this site Mickey Mouse but don't turn them in to Disney, OK? The official name is Tunnel View Arkansas Headwaters Recreation Area Site so maybe "Mickey Mouse" is just less of a mouthful!

This area is unclaimable due to being withdrawn in 1894 for a planned reservoir, which has yet to be built as mentioned in the previous site description.

Local Hints & Cautions:
- The magic here is the dispersed camping sites on this piece of BLM land. You will see about 9 easy river access points, each with a spot to set up camp. If you are visiting for the day, avoid "invading" the space of active campsites without asking first.

Gold Finding Tips:
- The prospecting here is fairly typical for the upper Arkansas River -fine gold for the most part, better where the riverbed material has some stickiness to it

Getting There: From the intersection of US-24 and CR-384 at 38.8760, -106.1594, turn east onto CR-384 and then immediately take the right onto Elephant Rock Lane. Follow this road past the end of the pavement to signage on the left which mentions "Tunnel View" at 38.8685, -106.1474. Take the hard left turn to go north on Tunnel View which drops down a steep hill and then turns east to a fork in the road with porta-potties. At this point the road is on BLM land. Head north or south at the fork and you will notice a combined total of about 9 easy river access points to your east.

This area can also be accessed from the other side of the river although access is more challenging. Take CR-371 north from BV to parking at 38.8744, -106.1446 which is a couple hundred feet off of CR-371 to the east.

Locale: just north of Buena Vista in Chaffee County
Land Type: medium sized river in a mountain valley
Land Manager: BLM

Boundaries: The upstream edge of this area is a private property boundary at 38.8815, -106.1501. The downstream edge of the prospecting access is 38.8684, -106.1425.

Key Regulations:
- NO digging into the slope of the banks on the east side or where rock has been added to stabilize the banks and to prevent erosion.
- A permit is required to use power equipment of any sort on the Arkansas River here. The Arkansas Headwaters Recreation Area office in Salida provides the permits.

Nearby Attractions & Accommodations:
See prior site.

The Arkansas River in central Pueblo

CHAPTER H: BUENA VISTA & THE LOWER ARKANSAS RIVER

The Arkansas River is known for large amounts of very small gold particles. It's also known for having surprisingly good amounts of gold fairly far downstream, over one hundred miles from the original hard rock sources. How is this possible? Well, large glaciers formed in the upper reaches of the Arkansas River during the ice ages. Huge lakes formed as the glaciers melted. Then when ice and rock dams broke, the lakes dumped out into the river and giant floods occurred. These events caused large amounts of gold bearing material from the glacial moraines to be blown downstream in torrents of water. These floods created the shape of the river as we see it today, and dropped gold deposits down on the plains where the energy of the floods dissipated. This means some spots around Pueblo and beyond have gold. The incredible floods also pushed placer gold further up onto hillsides and benches so some creeks feeding into the Arkansas River have gold but only in the lower part of the creek.

Driving US-50 along the Arkansas River from Kansas to Pueblo CO, you are on a major gold rush migration route and a part of the famed Santa Fe Trail. The lowlands along the river valley were also crucial property for the plains Indians who needed the forage, wood and shelter provided by the valley in winter. Here too gold has accumulated in the river gravels hundreds of miles downstream of the source deposits in the mountains. It's rarely accessible unless you are inspired to knock on the door of a rancher whose place backs up to the river to ask for permission. However, I found one spot and of course a stop at the re-created Bents Fort will provide a fun diversion from the road and a quick education on the importance of this route in our nation's history-before, during, and after the Pike's Peak Gold Rush.

Chalk Creek

In the first guidebook, I included four sites on Chalk creek. A more thorough review of the area for this book produced a couple more. Both of these are downstream enough that they are a short side trip for those traveling US-285.

Site Number: H-02E
Site Name: Chalk Creek Interpretive Site

Just downstream of the Mount Princeton Hot Springs resort is this interpretive site which tells of the local history. This is also an access point to the Colorado Trail, which reaches all the way from the west edge of metro Denver to Durango in the far southwest of the state.

Local Hints & Cautions:
* This is a busy site so be prepared for curious tourists and even kids who want to learn to pan. At least access is easy.
* You are being watched and judged, be a good ambassador of our passion by keeping your tailings smoothed out below water level and filling any holes.

Gold Finding Tips:
* Dig through the loose material in the upper few inches of creek bed to get to the dense, hard packed true creek bed where the gold is waiting.

Getting There: This site is on CR-162 just downstream of the Mount Princeton Hot Springs on the south side of the road and about six miles west of US-285. There is a parking area and a bridge over the creek so access to either side of the creek is easy.

Locale: Buena Vista
Land Type: medium creek
Land Manager: USFS

Boundaries: Upstream: CR-162 and its culvert. Downstream: Do not dig in the pond so the border is the boundary between the moving stream and the still pond.

Key Regulations:
- Pans only
- No digging in the banks, on dry land or in the pond. Stay upstream in the flowing creek area.

Nearby Attractions & Accommodations:
The town of Buena Vista and the hot springs resort.

Site Number: H-02F
Site Name: Old Chalk Creek Highway Bridge

Just before Chalk Creek runs under US-285, it runs under an older bridge that is where the highway used to run back in the day. A well-marked stretch of the creek here is accessible to us and it has some of the better gold I have seen on Chalk Creek. Perhaps the gold is a combination of material flowing down the creek and material from the historic floodplain of the Arkansas River. We are that close to the river at this site.

Local Hints & Cautions:
- A quirky little out of the way spot right next to the main highway but you will go unnoticed by the many people driving right by; and yet very easy to access.

Gold Finding Tips:
- There is an artificial drop in the creek left by road construction (between the old bridge and the new one). I quickly found color there.

- With a little effort, a sluice could be set up effectively here.

Getting There: This site is on CR-162, just barely upstream of US-285. There is an informal parking area just north of the bridge over the creek so access to either side of the creek is easy. To get there easily, turn off of US-285 onto CR-162 and look for a turn to the right almost immediately. The road drops downhill a little to the bridge and right past that to the dirt parking area north of the bridge. You are headed to 38.7414, -106.0830.

Locale: Buena Vista
Land Type: medium creek
Land Manager: CDOT

Boundaries: Upstream: Just about 50 feet upstream of the old bridge you will see a fence marking the edge of the private property. Downstream: Just downstream of the US-285 bridge, private property starts again. It is marked.

Key Regulations:
- No gas-powered gear here since the whole site is within 50 feet of CDOT structures (bridges).
- No digging where you would be disturbing CDOT erosion control structures, etc.

Nearby Attractions & Accommodations:
The town of Buena Vista and the hot springs resort.

The South Fork of the Arkansas River
I never noticed there was a South Fork of the Arkansas River until I started exploring Colorado with my gold pan! The South Fork of the Arkansas River stretches from the east side of Monarch Pass, down along Highway 50 toward US-285 near Salida. This stretch of US-50 is known as The Ghost Highway because the state highway department built the road right on top of several pioneer cemeteries without even bothering to move them!

While there was never a gold rush to this area in the traditional sense, the area did boom for a time. The historic highlights of this area are the Madonna Mine and the ghost town of Arborville.

The Madonna Mine, up fairly close to the pass, was a major producer of several metals including gold. It remains in operation today although their primary modern-day product is roadbed gravel produced from the old tailings piles. There are a couple of pull-offs on the side of US-50 where travelers can stop to view the huge, hillside mine and take pictures. Uphill, above the mine, there is a USFS campground called Monarch Park. While panning is allowed there, you won't find anything since the mineralized areas are all downstream. Oh well.

A bit downhill of the mine is the Monarch Mountain Lodge and across the street from it is a ski rental shop next to a 4-wheel drive road (FR-230). A drive up this relatively challenging road, along the middle fork of the South Arkansas River will take you past three old mines – you will see the mine dumps on the right side of the road as you travel up hill. Those who drive to the very end of the road and then walk up the trail a few hundred yards will see another mine dump. The remains of an old cabin crown this dump and a stream runs out of the collapsed mine portal.

Downstream from the mines, it is possible to find a little bit of gold in the river. The best place to access this legally is the Monarch Spur RV Park. However, just before we get downhill to the RV park, we need to mention Arborville. The remains of this ghost town are close to the highway but well below it due to the steep hillside. Within just feet of the highway, and most visible, is the 100+ year old stage stop building. It is a three-story building with a dark mansard roof. One of the very first concrete buildings in the region, it was first a stagecoach stop and pub, giving travelers a rest stop on the trip from Salida to Gunnison. Later, when the mines opened, a bordello took over the building. Other remaining structures include two log cabins and various smaller outbuildings. The population of Arborville peaked around 150 in the 1880s with many residents working in the mines just uphill.

> **Site Number: H-08**
> **Site Name: Monarch Spur RV Park**

This charming RV Park and campground is built on an old placer mine site. The site was first developed into a campground in 1960 but has recently been extensively renovated. The original owner did a little panning in the river now and then, according to stories passed down from owner to owner.

Local Hints & Cautions:
* The prior owners suggested panning in the river at the playground. This interesting area is just below a small waterfall created by a giant culvert carrying the river under the highway.

Gold Finding Tips:
* The gold here is, in the words of the owners I spoke with in 2019, "minuscule". I have to agree! I did find a few, very small colors in my sample pans, but candidly not much. I almost needed my reading glasses just to see the tiny specks. This would be a fun place to pan if you like a challenge and your kids are enjoying the playground equipment when they aren't "helping" you gold pan!

Getting There: The Monarch Spur RV Park is on the south side of US-50 11.7 miles west of Salida on US-50 between mile markers 208 & 209 at 18989 West US-50; call for reservations: 719-530-0341.

Locale: Western Chaffee County, a bit east of Monarch Ski Area and Monarch Pass
Land Type: small river running through a narrow mountain valley
Land Manager: private property

Boundaries: Upstream: the highway culvert. Downstream, the east edge of the RV park (if the campsite next to the river is occupied, be sure to ask the occupants for permission before walking through their campsite).

Key Regulations:
- Only open to those staying at the RV park.
- Ask permission, ownership has just changed as of 2023.

Nearby Attractions & Accommodations:
If you are digging here, you are staying here. They accept everything from tents to full-sized RV's up to 60 feet long and provide full hookups. While here, take a look at the Arborville ghost town just uphill from the RV park. Another fun thing to do in this area is to visit the Monarch Crest store, restaurant and scenic chairlift ride. The views from the top of the 7-minute chairlift are amazing as you are on top of the continental divide in one of the most picturesque parts of the US Rockies.

Site Number: H-09
Site Name: McPhelemy Park

Cottonwood Creek runs through this park which is right off of US-24 in Buena Vista. During the summer, the dam at the downstream end of the park is closed and a large pond is created.

Local Hints & Cautions:
- As with many city parks, the rules are pretty strict here. Pans only and no digging into the banks or on dry land in any way. You are being watched so be a good ambassador for our passion!

Gold Finding Tips:
- The up-steam area at the north end of the park is the place to focus.
- The gold here may have been deposited by the mega-floods of the Arkansas River during the ice ages or also moved downstream by Cottonwood Creek. I tried sampling other public access spots a bit upstream and came up with empty

pans. This makes me speculate that the gold is from the Arkansas River.

Getting There: This site is on US-24, on the west side of the road at the intersection of Main St and US-24 in Buena Vista.

Locale: Buena Vista
Land Type: small creek in a city park
Land Manager: Town of Buena Vista

Boundaries: Upstream: The road and its culvert. Downstream: Do not dig in the pond so the border is the boundary between the moving stream and the still pond.

Key Regulations:
* Pans only
* No digging in the banks, on dry land or in the pond. Stay upstream in the flowing creek area.

Nearby Attractions & Accommodations:
The town of Buena Vista, the nearby Arkansas River with its many prospecting opportunities. There is an ice cream shop just across the street!

| **Site Number: H-01BB, H-01CC** |
| **Site Name: Texas Creek** |

Texas Creek is a tiny little community at the junction of CO-69 and US-50 in the Arkansas River valley. This site fits between H-01Q and H-01R in the first guidebook.

The Texas Creek BLM site is a bit unique – both parking and access are wedged in around private property. The north parking is also used by ATV enthusiasts since it is the starting point of some extensive ATV trails to the north. Whether you use the north parking site or the west one, please be sure to only park in a designated area. Both parking areas are quite close to the dig area. There's only public access to the river here because the private landowner generously allows the public passage across their land. Be a good guest: stay on the established paths and only dig on the strip of public land.

Local Hints & Cautions:
- This site didn't make it into my prospecting guidebook, *Finding Gold In Colorado: Prospector's Edition*, because the access across the private property was unresolved. Thanks to the BLM and the landowner for their support of prospecting! Please leave this dig site as you find it (or better!) so we continue to have access.
- During rafting season, the water here is really too high for much prospecting. This is a better dig site from August 16th into autumn. The site is south facing, in a canyon, and low enough in elevation to stay warm well into the fall so enjoy!

Gold Finding Tips:
- The inside bend here, combined with the turbulence caused by the confluence with Texas Creek, makes this a productive panning site.

Getting There:
There is a BLM-provided map of this spot on my website which may help visitors understand the parking and access.

H-01-BB: This site is 27 miles west of Cañon City on US-50 in Bighorn Sheep Canyon (Yes, there really are bighorns here!)

Turn off of US-50 at CR-27 (38.4097, -105.5845) and cross the bridge. The bridge over the river is just west of the intersection of US-50 and CO-69. You'll also see rafting company buildings on the highway side of the river. As mentioned above, there are two parking areas. The first is just up CR-27 where there is a fenced off parking area on the right at 38.4154, -105.5856. That one can get filled with trucks and ATV trailers, so it is more thoughtful and more convenient for us prospectors to go on to the second site which is also closer to the river. To get there, instead of entering the first parking lot, take the left turn across the dry wash and back toward the river. The road loops back south to a lot at 38.4119, -105.5860. From there just walk east across the dry wash to the road and then south to the river to access the prospecting zone. See below for details.

H-01CC: The other unclaimable patch of land here is accessed from the highway side of the river. The pull-off is just downstream of site H-01BB and around a bend. Turn off of US-50 at 38.41739, -105.5746 where there are a couple of dirt loops and parking between the highway and the river. The parking is on a gentle inside bend, so the dig area is spitting distance from the parking.

Boundaries:
Site H-01BB: The prospecting area is just east of the road (CR-27) bridge. Please respect private property by not digging further upstream; even the bridge itself is on private property. Legal access starts at 38.4107, -105.5827 and extends around the long inside bend to 38.4117, -105.5779. Access includes the island and both banks of the river.

Site H01CC: Access starts a bit upstream of the informal parking at 38.4128, -105.5760 and extends north, downstream to 38.4196, -105.5745.

Locale: Arkansas River Canyon between Salida and Canon City
Land Type: river running through a narrow valley
Land Manager: BLM

Key Regulations:
• No restrictions except a warning not to damage woody plants.

Nearby Attractions & Accommodations:
There's not much here so come prepared to take care of yourself!

Site Number: H-01DD
Site Name: Blue Heron Open Space

Blue Heron Park is on the Arkansas River just east of Florence. This large park and open space area is just downstream of H-06 in the first guidebook. The main uses of this park are fishing, including in the small lakes, bird watching and boating. Let's add gold prospecting to that list!

Local Hints & Cautions:
• See points included in the prior site.

Gold Finding Tips:
• Typical Arkansas River gold is reported but your author has not sampled this specific area.
• The river is braided here so expect multiple islands and flows of water, not just one.

Getting There: Access on north side of river from CR-115 to Blue Heron Park Rd to parking at 38.3959, -105.0636. Plan on a ¼ mile walk past the fishing ponds to the river.

Boundaries: Upstream access starts at the CR-115, aka Vietnam Veterans Memorial Hwy, bridge over the river at 38.3904, -105.0679. Downstream is basically due east of the parking area at 38.3958, -105.0563.

Locale: just east of Florence
Land Type: moderate braided river in a wide valley
Land Manager: BLM, Arkansas Headwaters Rec Area

Key Regulations:
• No powered equipment without a permit from the Arkansas Headwaters Rec District (office in Salida)

Nearby Attractions & Accommodations:
The town of Florence is just west. There are fun restaurants and the town is known for its antique shops.

Site Number: H-07C
Site Name: East Edge of Lake Pueblo State Park

The eastern edge of the state park has seen multiple uses through the years; you'll see remnants of this history at this site. This site is just downstream of H-07B in the prior guidebook.

Local Hints & Cautions:
• Land ownership is mixed here between state park, city/county and private, so be careful to prospect where you are welcome.

Gold Finding Tips:
• Check all the usual spots. As far as I can see, this area is not well used by local prospectors, so there may be pockets of better gold in fairly 'obvious' spots around larger rocks and little bends in the river.

Getting There: From CO-96 on the west edge of Pueblo, go to the Volco Ponds & Arkansas River Access park at 38.2612, -104.7018. The road is open from late May through late November.

Locale: Pueblo County between the city of Pueblo and the dam at the eastern edge of Pueblo Lake State Park
Land Type: moderate river running across a plain
Land Manager: State Park land and city/county property

Boundaries: Upstream: The river at the parking area. Downstream: Valco Bridge, then river access on the north side of the river only toward town, merging into the next dig site.

Key Regulations:
• Prospecting rules in state parks are very, very strict. Pans only and only take home what fits in your pocket. Review the details in the introduction to this book.
• Only dig IN the river.

Nearby Attractions & Accommodations:
The state park itself of course. The town of Pueblo with its Riverwalk, local breweries, and El Pueblo History Museum (a History Colorado museum).

Site Number: H-10
Site Name: Runyon Lake Park

Runyon Lake is a large park in central Pueblo with lots of facilities including the lake and a whole sports complex. The bulk of this park is on the north side of the river, but we are going to park on the south side since all we care about is the river, right?!?

Local Hints & Cautions:
- This is an urban area so assume there is trash in the river – wear gloves and bring a trash bag to collect whatever is in your immediate area.
- Park on the south bank of the river as directed in the driving instructions to dramatically reduce how far you have to walk.
- The pedestrian bridge over the river here means access to both the north and south banks is easy. The bridge also makes a great vantage point to survey the river to see where you want to dig.

Gold Finding Tips:
- Do all the usual stuff, sample around, check exposed cobble bars, focus on inside bends.
- It's fairly easy to find at least a little gold here but I've never seen a lot. Maybe you will be luckier, some folks are!
- This area is fairly popular with local prospectors who pan and sluice successfully but is still very lightly used.

Getting There: From exit 98A on I-25, turn left onto Santa Fe Ave. In just a few hundred feet you will cross the bridge over the Arkansas River (resist the urge to gawk!) and then bear left onto Santa Fe Drive. Again, almost immediately, turn left onto

Moffat St. and follow it to the end of the road where you will find a parking lot on the right at 38.2531 -104.6054. Now walk north on the path toward the river, it's only a few yards.

Locale: central Pueblo, just east of I-25
Land Type: larger river in an urban area
Land Manager: City of Pueblo

Boundaries: Upstream: the dam. Downstream: the confluence with Fountain Creek (coming in from the north) with access continuing downstream from the north shore only to 38.2528, -104.5852.

Key Regulations:
• No digging in the vegetated banks.
• No gas-powered equipment.

Nearby Attractions & Accommodations:
Pueblo has some fun and quirky things to do such as visiting Walter's Brewery, touring the history museum, the historic Pueblo Union (train) Depot, the museum at the Steel & Iron Works and exploring the old downtown Riverwalk district. Be sure to check out Neon Alley at night.

Site Number: H-11
Site Name: Arkansas River Trailhead

The trailhead here has its own parking and provides access to the confluence of the Arkansas River and Fountain Creek from north side of the river right in the middle of town.

Local Hints & Cautions:
• Land ownership is mixed here between state park, city/county and private so be careful to prospect where you are welcome. You will be fine within the flood zone of the river, just don't wander around on the land above the river banks.

Gold Finding Tips:

- Check all the usual spots. As far as I can see, this area is not well used by local prospectors so there may be pockets of better gold in fairly 'obvious' spots around larger rocks and little bends in the river.

Getting There: Near the intersection of South La Crosse Ave. (CO-227) & Stockyard Road on the west side of South La Crosse at 38.2561, -104.5887 Pull into the parking lot and walk west to the river.

Locale: The city of Pueblo
Land Type: moderate urban river running across a plain
Land Manager: city and county property

Boundaries: Not relevant as long as you stay within the flood zone of the river.

Key Regulations:
- Only dig IN the flood zone of the river.

Nearby Attractions & Accommodations:
The state park west of town. The town of Pueblo with its Riverwalk, local breweries, and El Pueblo History Museum (a History Colorado museum).

Site Number: H-12
Site Name: La Junta River Access

The city of La Junta owns some Arkansas Riverfront property adjacent to the Main St. bridge over the river. This gives us access to both sides of the river. I was excited to learn about this spot from a local prospector because it is so far out on the eastern plains.

Local Hints & Cautions:
- These access points are multi-use. Be sure to park in a way that allows people with a truck and trailer to maneuver easily.

- The parking areas are fairly informal. Only park where others have in the past and where there is no vegetation.

Gold Finding Tips:
- Expect the gold to be small. I haven't sampled this area.
- The river is somewhat braided here so explore the sand/cobble bars in mid-river if the water is low, as it often is this far east.

Getting There: From US-50, turn right onto Brandish Ave., then left on 3rd St. then left to turn north on Adams/Main St. (CR-109). Take the bridge up over the railroad tracks to:

H-12A: Exit west from the bridge onto North St. to drop down onto the south bank of the river. Follow your nose a bit to go back under the bridge to park on the city property just east of the bridge. From there consider walking upstream to the big inside bend there.

H-12B: Continue north on the bridge to the far bank of the river. Take the first left onto Vine St. and then another left almost immediately to head toward the river and informal parking near the bridge. Consider walking downstream to the inside bend. If conditions are safe, it may also be possible to drive downstream under the bridge to closer parking.

Boundaries:
There are no practical boundaries as long as you stay within the waterway. Do not go up on dry land where you may be trespassing.

Locale: La Junta, on the eastern plains about 60 miles east of Pueblo
Land Type: river running across the plains through a town
Land Manager: City of La Junta and Otero County

Key Regulations:
- None noted. Follow all posted signage.

Nearby Attractions & Accommodations:

Stay and eat in town or at a nearby campground. Be sure to visit Bent's Old Fort Historic site a bit east of town along US-50.

CHAPTER I: CRIPPLE CREEK

"Crazy Bob" Womack was the tenacious prospector who discovered the riches of the Cripple Creek area. The whole story is in the prior guidebook. When you visit, be sure to hike over to see Bob Womack's cabin. When I wrote the prior book, I didn't even know it still existed! The cabin is near the train station in Cripple Creek. It's a short hike over the hill. To find it from the train station, walk east on Carr Ave. for a block, take the right turn onto Main St. Follow that south a half block to the trail which goes further south. Walk the trail around and down the hill to the cabin at 38.7460, -105.1696; have a good look but don't go in or attempt to take anything as a souvenir.

Prospecting Opportunities
A lot of the land here is owned by the Cripple Creek & Victor Mining Company so finding a legal, unclaimable spot to dig is tricky. In the first guidebook, I covered the creek flowing south out of the town of Cripple Creek, this time I was able to add one site on the creek flowing south from Victor.

Site Number: I-02A,B
Site Name: Wilson Creek in Victor

Wilson Creek runs south along the east edge of Victor, past a lot of old mines. It then turns west along the south edge of town before heading south again to flow toward the Arkansas River, many miles away. This site is on the south edge of town so access is easy.

Local Hints & Cautions:
• The creek is often dry once the snow melt is done.

Gold Finding Tips:
• This tiny creek is fun to explore. The gold is patchy in my

experience.
- Dry washing may be more practical than panning since the creek is often dry by July or so. Alternatively, learn to dry pan or classify some dry material and take it away to pan.

Getting There: From downtown Victor, follow 2nd St south to the edge of town. It curves around to the east and becomes Lewis Ave. but there's no signage to indicate this. Parking along the side of the road should be fine or park at 38.7017, -105.1463 where there is a wide spot just before the city maintenance yard. Be sure to avoid blocking the road as maintenance vehicles use it. Also, there are residences northwest of the maintenance yard which use this road as their only access.

Boundaries:
Since the creek is running from east to west here, keeping an eye on the longitude is enough to keep you legal.

I-02A: Upstream from 38.701, -105.139 to 38.701, -105.142.

I-02B: Upstream from 38.702, -105.146 to 38.701, -105.148 behind the city maintenance yard.

There is private property all around so please respect these boundaries even though all of the land along here looks open.

Locale: on the south edge of the little town of Victor
Land Type: small creek in a valley
Land Manager: Town of Victor

Key Regulations:
- Respect any signage along the road or on-site.

Nearby Attractions & Accommodations:
As I shared in *Finding Gold in Colorado: Prospector's Edition*, I just love the town of Victor. It is absolutely the best-preserved mining town left in Colorado without all the influences of skiing, tourism and such that burden other towns.

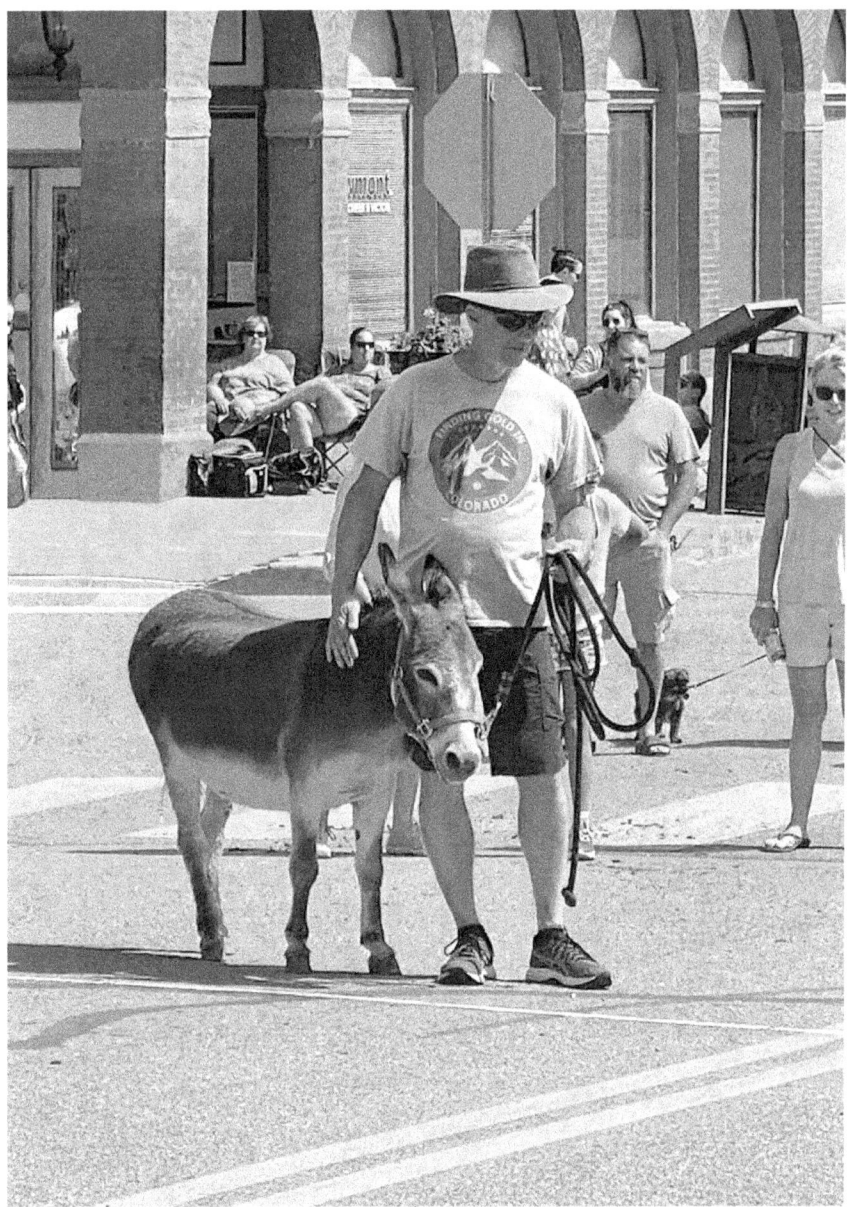

Ready for the 2021 mini burro race in Victor; my little buddy
and I won the race!

CHAPTER J: TAYLOR PARK & PITKIN

The little town of Pitkin and its even smaller neighbor, Ohio City, are all that remain of a local gold and silver rush in the late 1870s and early 1880s. This area is accessed by turning off of US-50 at Parlin, which is about halfway between Monarch Pass and Gunnison. County Road 76 runs up along the Quartz Creek valley following the old route of the Denver, South Park & Pacific Railway. The rails are long gone but visitors can still see the old railroad station in Pitkin. There are quite a few other older buildings here too, sprinkled in amongst newer summer homes and shops. Look for the town hall, community church, assay office, telegraph office, jail, mill, and museum in the old schoolhouse, as well as a wide variety of well-preserved private homes. Ohio City, similarly, has an interesting old town hall, jail and schoolhouse along with a mix of well-preserved older homes and newer ones.

The first miners arrived in 1877 and made significant discoveries of silver and gold in this area in 1878. The town was started in early 1879 as Quartzville by Frank Curtiss, George Chiles, and Wayne Scott. By August, they had succeeded in incorporating the town and renaming it Pitkin, in honor of the then Governor Fredrick Pitkin. By 1880, the boom was on and the area census that year showed 1800 residents. Unlike many mining boom towns, many prospectors and miners came with their families, clearly planning to stay. Both a church and a school were built that year with 39 children attending in the fall of 1880. This stands in sharp contrast to many other boom towns where there were virtually no children in the early years and any church service was held either in an open field or in a local saloon.

Sadly, folks guessed wrong when they imagined this town would thrive. The population probably peaked in 1882 when there were over 60 businesses and 300-400 homes in town with many more in the nearby hills. Indeed, many of the mines were very successful at first. However, as the mines reached a depth of 75-100 feet down, the gold and silver pinched out in the veins and dreams of quick riches or long mining careers faded. By 1883, just four years after the initial boom, the population had dropped to less than 500 people. Similarly, the train came to town in 1882 but after just a couple good years, the rail line use declined. The last trains were running until 1911 but perhaps more because of traffic to Gunnison than for people wanting to get to little Pitkin.

Other challenges for this mining district in the 1880s included three major town fires, three epidemics and of course the termination of guaranteed prices for silver by the US government. When that happened, in 1893, the so-called Silver Panic occurred and the market price for silver dropped over 70%. Any mine which lacked decent gold production, and depended on silver, simply shut down and sent all of their workers home. Today there are less than 100 full-time residents in Pitkin and even fewer in Ohio City. Both towns have an influx of summer "residents" who enjoy four-wheeling and hanging out at the Silver Plume General Store and Grill. There are, however, several mines still operating along Gold Creek above Ohio City. Learn more about the Pitkin area and its history at www.pitkincolorado.com.

The School House Museum in Pitkin is worth a visit since they cover quite a bit of the mining history (along with railroading, schooling, etc.). You can find the museum on Main Street east of 6th Street. Look for the white clapboard building on the south side of the street (and ignore the private building at 3rd street which says it is a museum, oddly enough it is not).

Ohio City has an 1897 schoolhouse, and both a jail and city hall dating back to 1906. There are a variety of other interesting, old buildings on both sides of the highway.

Site Number: J-04
Site Name: Quartz Creek Campground

This stretch of Quartz Creek, outside of the town of Pitkin runs through an area set aside as a recreation area by the USFS to build the Quartz Creek Campground. The campground is fairly primitive, lacking hookups or a proper dump station, but it does have pit toilets, and some people are brave enough to pull their travel trailers up here. Actually, the road is quite good since the county grades it every summer; the road through the campground is worse, so just take it very slowly and think twice if you have a budget trailer with a flimsy frame! This whole area is very popular with the ATV/UTV crowd, so it fills up on the weekends in summer. If you show up during the week, you will have your choice of spots.

Local Hints & Cautions:
* Respect the sensibilities of your fellow campground users. Ask permission to walk through campsites to access the creek in the campground.

Gold Finding Tips:
* The gold here is typically extremely small and deposits are spotty. Sample around to find better concentrations.

Getting There: From US-50 at Parlin, head northerly and easterly on CR-76 to Pitkin. Continue through town and follow the signage directing you toward Cumberland Pass. The campground is several miles out of town on FS-765.

Locale: rural Gunnison County
Land Type: medium sized creek running through a mountain valley
Land Manager: USFS

Boundaries: Upstream at 38.6416 -106.4698 to downstream at 38.6361 -106.4699.

Key Regulations:
- Do not disturb woody plants.
- There is a day use fee if you access Quartz Creek from the campground. However, there is no fee if you access it from the road which follows fairly close to the creek. There are places to park along the road near the creek. Folks seem quite tolerant of trucks pulled off on the side of the road but be considerate of other's needs.

Nearby Attractions & Accommodations:
The towns of Ohio City and Pitkin of course. Also check out the driving tour of Cumberland Pass below since that is on the same road and just a bit to the north.

Drive Tour Over Cumberland Pass

Driving further up the FS-765 road from the Quartz campground, the route goes all the way up to 12,000+ feet at Cumberland Pass and then down the other side to Tin Cup (see my first guidebook for info on Tin Cup). There are some beautiful views from the pass and along the way, an impressive mine ruin to visit. At 38.6822, -106.4810, you will see several large miners' cabins and on the hill above them, a large mine dump with the ruins of the large ore processing building. The actual portal has collapsed but there is still water running out of it. I found some interesting mineral specimens here, maybe you will too.

Site Number: J-05
Site Name: Pitkin Campground

This stretch of Quartz Creek, just east of the town of Pitkin runs through an area set aside as a recreation area by the USFS to build the Pitkin Campground. The campground is fairly primitive, lacking hookups or a proper dump station, but it does accommodate larger RVs fairly easily. This campground is very popular with the ATV/UTV crowd, so it fills up on the weekends. The campground is walking distance from town.

Local Hints & Cautions:
- Respect the sensibilities of your fellow campground users. Ask permission to walk through campsites to access the creek in the campground. The easiest access in the campground is from sites 10-14 and near the bridge over the creek at the entrance to the campground.

Gold Finding Tips:
- The gold here is typically extremely small and deposits are spotty. Sample around to find better concentrations.

Getting There: From US-50 at Parlin, head northerly and easterly on CR-76 to Pitkin. Continue through town to the campground on the east end of town just past the end of the pavement.

Locale: Rural Gunnison County
Land Type: medium sized creek in a mountain valley
Land Manager: USFS

Boundaries: Upstream at 38.6124, -106.4930 to 38.6116, -106.5014. Since the creek runs basically east to west, simply watching your longitude location will suffice here.

Key Regulations:
- Do not disturb woody plants.
- There is a day use fee if you access Quartz Creek from the campground on the south side of the creek. However, there is no fee if you access it from the road which follows the creek fairly closely on the other side of the creek. There are places to park and apparently even to do free, dispersed camping on the north side of the creek.

Nearby Attractions & Accommodations:
The towns of Ohio City and Pitkin of course. Also check out the driving tour in this chapter. This campground is a quick walk from the general store in Pitkin (which has free Wi-Fi and LOTS of t-shirts).

Downstream from the town of Pitkin, the Roosevelt Picnic Area (no actual tables or other facilities) offers another chance to prospect Quartz Creek. There is also an excellent, short hike near here to the old Roosevelt Mine, mill and power station.

Site Number: J-06
Site Name: Roosevelt Picnic Area

This stretch of Quartz Creek, two miles west of the town of Pitkin runs through an area set aside as a recreation area by the USFS to preserve the remnants of the Roosevelt Mine and to provide fishing (and prospecting) access to the creek.

Local Hints & Cautions:
- Be sure to park fully off the road and in such a way that others can use the limited parking.
- Creek access is fairly easy but some agility is needed here.

Gold Finding Tips:
- The gold here is typically small and deposits can seem spotty. Sample around to find better concentrations.

Getting There: From Pitkin city limits signage, travel about two miles west on the main highway (CO-76). There are spots to park near either end of the area but not really anywhere safe to park in between.
- The first parking spot is on the south side of the road at 38.5833 -106.5387 which is slightly upstream of the unclaimable area - so using this spot means walking downstream a bit. The good news is you'll find plenty of parking here and you could even spend the night.
- The second parking spot is on the north side of the road at the downstream end of the public access area. Here just turn off of the highway onto the old road which is open for about a hundred yards before being blocked from there on. Spending the night here would be an option.

Locale: Rural Gunnison County
Land Type: medium sized creek running through a mountain

valley
Land Manager: USFS

Boundaries: Upstream at 38.5818 -106.5401 to 38.5783 -106.5425. Since the creek runs basically east to west, simply watching your longitude location will suffice here.

Key Regulations:
- Do not disturb woody plants.
- Do not disturb any structures or anti-erosion features related to the roadway.
- Do not remove or disturb anything related to the historic mining operation – not even 'trash'.

Nearby Attractions & Accommodations:
The towns of Ohio City and Pitkin of course. However, the highlight of this spot is the Roosevelt Mine complex. The remaining buildings include a concrete structure that originally housed an electrical generation plant, remains of a water system to feed the power plant and the mill above it, the relatively intact stamp mill, a large mine dump, and an interesting stone and concrete mine portal building including the preserved portal.

Roosevelt Mine Hike

The Roosevelt Mine complex is accessed by hiking only but it's a fairly short walk – figure on less than a mile out and another back. To get to the hiking access, drive about ¾ mile west from the downstream end of the public access area to a wide parking area on the south side of the road just west of mile marker 12. Follow the "Fisherman Access" trail from the parking area, cross the bridge and take a left. Follow this narrow, minimally used path upstream along the creek to the concrete structure of the old electrical generation plant building. The plant produced power for the mine. From here the trail can be confusing but follow your nose upstream and uphill to find the rest of the historic structures. Here is the best description of how to walk this loop:
- From the power plant, follow the narrow trail along the

water flume that supplied the plant. Along the path, you will see many of the steel bands which held the wooden flume pipe together.

- After passing just above the flume trestle bridge on this trail, it is time to transition onto the old mine road. To do this, walk and scramble up hill (you will see an informal, unmarked trail) about 20 feet to the old, graded mine road.
- Turn right and walk uphill on the mine road to the mine portal.
- After examining the portal and the remains of the building (notice the concrete bases which supported the boiler and air pump), follow the informal trail east to the mine dump. Explore the mine dump for interesting rocks and ore samples.
- Follow the informal trails downhill and to your right to the mill. This amazingly complete structure has fallen down to a partial degree but still holds its stamps, cams, and many other aspects of its operational configuration. I think it is the best example of a stamp mill that I have seen still standing, unpreserved, where it was last used.
- From here you can retrace your steps or simply follow the informal trails back downstream more or less along the old flume.

Ohio City

Ohio City is on the highway 6 miles west of Pitkin, so you drove by it to get to the town of Pitkin. Ohio City just barely still exists but it has quite a history of its own. The town was founded in the early 1860s as a gold mining town, but the boom didn't last, and the town failed. A silver boom reignited the town and it thrived again from 1879 to 1893 when the price of silver collapsed, and the town took its second dive toward ghost town status. But then in 1896, prospectors found the lode deposits that were the source of the placer gold which had been mined in the 1860s. This new success woke the town back up and it thrived until 1916 when the quality of the remaining ore fell to the point where it was no longer economic to mine here. A semi-ghost town again, Ohio City has stayed quiet since. These days it includes some summer homes and a store that is open seasonally. There are homes for rent here if that appeals.

A drive north, up CR-771, from Ohio City takes you up the little valley created by Gold Creek. There is indeed gold here in the creek of course. Along the drive up-valley, visitors will see old cabins and other mining related structures. There are also several active mines! Feel free to pull over for a closer look at the old structures on USFS land but stay off of the private properties with the operational mines. Happily, quite a bit can be seen from the road since the valley is so narrow. Visitors who drive up the road will eventually get past all signs of mining, and at about the 7-mile point, will arrive at Gold Creek Dispersed Campground. There are several marked sites here with picnic tables, a turn-around loop, and a toilet building. There are also quite a few other options along the upper part of the road for dispersed camping, even for those with travel trailers or other RV's. The road is not great but take your time and any RV can get up there. As you might guess, the fact that there is no evidence of mining on the upper part of the creek means there is no gold at the campground – disappointing but true. Happily, there are a couple of other areas in the gold bearing part of the creek which are unclaimable. Details on those are just below.,

As you drive back downstream toward Ohio City, make a note of all the old cabins at 38.6421, 106.5747 and the impressive old mill at 38.6314, -106.5799 which sits between two active mines on private property.

Site Number: J-07
Site Name: Gold Creek Recreation Area

This stretch of the Gold Creek valley was set aside as a recreation area by the USFS under PLO 2976 on 3/18/1963. While labeled a "picnic area" on some maps, there is no actual picnic area constructed. However, the designation does provide some protection to the historic mining structures in this area. Feel free to explore these old structures carefully and respectfully. Remove nothing of course.

Local Hints & Cautions:

- Be careful of hazards related to trash from the old mining operations. I found broken glass and sharp pieces of old metal in my sample pans. Other than that, access here is very easy.
- Be sure to park fully off of the roadway to allow other travelers to pass by, the road is only about 1½ lanes wide. There is a warning sign about parking completely off the road; you can expect the locals to call the sheriff if you disregard this restriction.

Gold Finding Tips:

- It is fairly easy to find at least a little gold in your pan here. Sample around to find better concentrations.
- Setting up a sluice should be fairly easy here as well.

Getting There: From US-50 at Parlin, head northerly on CR-76 to Ohio City. There, turn left on CR-771 and travel for several miles. Once you are within the recreational boundaries, there are plenty of places to pull off the road to prospect, many near old mining buildings. Here are the parking pull-offs I noted, listed from upstream to downstream, so you may want to drive all the way up to the campground sightseeing and then turn around to follow these instructions like we did:

38.6288, -106.5820,
38.6276, -106.5824 at an old cabin,
38.6263, -106.5826,
38.6240, -106.5838,
38.6233, -106.5842 by old mining buildings,
38.6222, -106.5847 by more mining buildings,
or just above or below the culvert at the downstream end of this prospecting area. <u>Do not prospect downstream of the culvert</u> as there are active claims in the next stretch of creek.

Locale: rural Gunnison County
Land Type: medium sized creek running through a narrow mountain valley
Land Manager: USFS

Boundaries: Upstream at 38.6307, -106.5808 to 38.6204, -106.5854. Since the creek runs basically north to south, simply watching your latitude location will suffice here. The downstream end of this access area is at a culvert where the stream runs under the road. Active mining claims and private property exist on either side of this access area so respect the boundaries I've provided.

Key Regulations:
- Do not disturb woody plants while you are digging.
- Do not disturb any structures, relics, or even 'trash' from old mining operations. Take no souvenirs except your gold concentrates.

Nearby Attractions & Accommodations:
The towns of Ohio City and Pitkin of course. Also check out the driving tour in this chapter.

Site Number: J-08
Site Name: Comanche Campground (closed)

This lower stretch of the Gold Creek valley was set aside as a recreation area by the USFS in order to build a campground. The campground is now closed but the area remains unclaimable so we can still visit and prospect.

Local Hints & Cautions:
- Be careful of hazards related to trash from the old mining operations. I found broken glass and sharp pieces of old metal in my sample pans. Other than that, access here is very easy.
- Be sure to park fully off of the roadway to allow other travelers to pass by, the road is only about 1½ lanes wide.

Gold Finding Tips:
- It is fairly easy to find at least a little gold in your pan here. Sample around to find better concentrations.

- As with the prior site, setting up a sluice should be fairly easy here.

Getting There: From US-50 at Parlin, head northerly on CR-76 to Ohio City. There, turn left on CR-771 and travel for several miles. Here are the pull-offs I noticed from upstream to downstream:
38.6029, -106.6000,
38.6017, -106.6012 and
38.5981, -106.6033; depending on your vehicle size, you may find other spots too.

Locale: rural Gunnison County
Land Type: medium sized creek running through a narrow mountain valley
Land Manager: USFS

Boundaries: Upstream at 38.6095 -106.5989 to 38.5938 -106.6051. Since the creek runs basically north to south, simply watching your latitude location will suffice here. Active mining claims and private property exist on either side of this access area so respect the boundaries.

Key Regulations:
- Do not disturb woody plants while you are digging.
- Do not disturb any structures, relics, or even 'trash' from old mining operations. Take no souvenirs except your gold concentrates.

Nearby Attractions & Accommodations:
The towns of Ohio City and Pitkin of course. Also check out the driving tour and hike in this chapter.

Old Adobe building, Town of Russell

CHAPTER K: SUMMITVILLE, CREEDE & THE RIO GRANDE BASIN

The entire San Luis Valley drains into the Rio Grande River so there is a lot of ground to cover here. The prior guidebook covered sites in the high country near Summitville and Creede. That is where the big rich mines were. This time, some of the smaller sites and more obscure locations are described. The San Luis Valley includes some of the most obscure gold prospecting areas in the state of Colorado. From rumors of lost gold mines in the hills on the eastern edge of the valley to traces of gold in many of the waterways across the valley, there are many hints of possibilities.

The first and most famous area is to the northeast of Fort Garland. Fort Garland was actually the second fort in this area. The first was Fort Massachusetts which was built in 1852 and abandoned in 1858 when Fort Garland was completed six miles to the south. While at Fort Massachusetts, the soldiers organized a mining company and placer mined nearby creeks during their off-duty time. This operation was several years before the 1859 Colorado Gold Rush and information about the successes of the soldiers may have contributed slightly to the gold rush. Spaniards and New Mexicans also placer mined in this area in the early to mid-1800s with enough success that their mining artifacts were discovered by those who came along in the 1850s. Even today there is a "Placer Creek" on the map here and a "Spanish Gulch" as well.

The Gold Rush Town of Russell

Readers of *Finding Gold in Colorado: Prospector's Edition* will recall that there are two areas mentioned in that book named for William Greenberry "Green" Russell. He led a group of prospectors to Colorado in 1858 and established the town of Russellville in Douglas County where there is still a road named Russellville. He left Colorado in late 1858, disappointed with his results at that point, but sure the gold was in Colorado somewhere. The next spring, he came back and found great gold deposits just over the ridge from Central City and Blackhawk. This area is still known as Russell Gulch. In early 1862, Russell was arrested by the U.S. Calvary. He was a Georgian and the US Calvary was charged with eliminating the confederate presence in Colorado. He agreed to leave the state and was allowed to bring his gold home with him. He used it to fund a Confederate Army calvary troop but apparently did not engage in any battles (he probably gave his word not to). Instead reports say he patrolled his home county looking for rebel army deserters.

After the Civil War amnesty was granted in 1868, Russell came back to Colorado and chose to prospect the area around what had been Fort Massachusetts. He was successful in discovering an alluvial plain loaded with gold and the town of Russell grew up. It was next to what is now US-160, which runs over La Veta Pass from the eastern side of the front range into the San Luis Valley. The ghost town's remains are just 36 miles west of Walsenburg (or east of Fort Garland) at 37.5553, -105.2878 where a boom-era adobe structure and several smaller buildings still stand. They are on private property but are easily visible from the road, near a CDOT facility.

In its day, the town of Russell was an exciting place. By 1876 the place had its own post office, always the sign of a town that had "made it". The main mining operations were on Placer Creek, Greyback Gulch and Willow Creek to the northwest and on "Big Hill" to the southeast but Russell remained at the center of it all. This was true, in part, because Russell was easy to get to from the eastern plains, so all of the stores, saloons, hotels and so on were established there. They even had multiple churches and proper schools. However, like most gold rush towns, Russell had its share of boom and bust. The first rush and the ensuing boom was from 1870 to 1871, then the town got quieter only to boom again when new discoveries were made in 1880. 1882 brought new discoveries just northwest on Willow Creek. Through the next years, a mix of placer mining, searches for the mother lode source, along with lumbering and iron ore mining kept the town humming. Then in 1898 the first dredge was brought into the area to process 800 cubic yards of gravel per day. In 1910, the much larger "Mary Blossom" electric dredge was brought in to dredge Placer Creek. It could handle 2,500 cubic yards daily but only operated for three years before the operators gave up and sold off the machinery. The shell of the dredge remains on the creek to this day – sad to say it is on private property with no way for the public to see it. In any case, when the big dredge shut down, it was the end of the town too. Locals tore down the buildings and moved them down to the Fort Garland area. A few hard rock mines continued operation into the 1930s but Russell was gone for good.

By the way, Green Russell himself left town in 1875. His son John had died in a mining accident in 1874, and, in 1875, the US government passed a law requiring all American Indians to live on a reservation or face imprisonment. Russell himself was part-Cherokee, but his wife was full-blooded. Rather than send her to the reservation alone, he went with her. He died on the Cherokee Reservation in what is now Oklahoma in 1877. He was a confederate rebel, and thus a traitor, but he played a big part in the gold rush stories of Colorado and stayed true to his family to the end. In 1914, the City of Denver created Russell Square Park in his honor, at 37th Ave. & Vine St. The park is still there today!

Sadly, for us, the creeks the soldiers mined, the area where Fort Massachusetts stood, and the whole area mined when the town of Russell was thriving, and the dredge, are all part of a very large private ranch now, so it is all off limits to the public. However, all of those creeks have to flow downhill somewhere, don't they? This book covers the waterway on public land that they all feed into.

> **Site Number: K-06A-C**
> **Site Name: Sangre de Cristo Creek**

Sangre de Christo Creek drains the area on the east side of the San Luis Valley where much of the mining associated with the towns of Russell and Fort Massachusetts occurred. Even though much of Sangre de Christo Creek is on private land, the state highway running through here includes a nice wide strip of land offering us multiple access points.

Local Hints & Cautions:
* Be sure to park well off of the highway for your own safety and that of others.

Gold Finding Tips:
* Avoid areas just above active beaver dams. A lot of sediment builds up in these areas making prospecting difficult.

Getting There:
K-06A: On CO-160, about 0.9 miles downhill from the remains of the town of Russell, the creek crosses under the highway giving us access on both sides of the road at 37.5434, -105.2863 with a decent spot to pull off the road.

K-06B: Continue downstream another 1.2 miles to 37.5287, -105.2970 with great parking on the north side of the road. I found color in every pan here.

K-06C: Another 0.9 miles downstream to a turnoff to the south onto Beekman Rd at 37.5157, -105.3035.

Boundaries: From the culvert under the road to the ranch fence in each case. In each case, it is at least fifty feet of creek, or more depending on the angle of the creek as it cuts though the state land next to the highway.

Locale: CO-160 between La Veta Pass and Fort Garland
Land Type: small creek
Land Manager: CDOT

Key Regulations:
- No power equipment ever here since all of it is too close to the state highway.

Nearby Attractions & Accommodations:
Check out Fort Garland Museum and historic site. Check out the food at All-Gon Restaurant at 319 Beaubien, right off of US-40 in Fort Garland. There are also lots of places to stay or find a decent meal in Alamosa.

Prospecting the West Side of the Valley
On the western side of the San Luis Valley, other waterways carry gold down toward the sands of the valley. These rivers include the Alamosa River and the Conejos River. Let's start by exploring the Alamosa River and then move further south to the Conejos.

Site Number: K-07
Site Name: Stunner Campground

Stunner USFS Campground is a very simple facility in a high mountain valley next to the Alamosa River.

Local Hints & Cautions:
- Avoid prospecting this area in April and May due to high water conditions.

Gold Finding Tips:
- I haven't explored this spot but the river has some nice inside bends to sample here. I have also heard good things from locals.

Getting There: From US-285 in downtown Alamosa, it is about

48 miles to the campground. Because of the forest service roads, budget two hours to get to Stunner Campground from Alamosa. The campground is on FR-380 at 37.3779, -106.5733; to get there from Alamosa, turn off of US-285 onto CR-10S/CO-370 west bound at 37.4296, -105.8912 and take that 14.1 miles to CO-15 south. Two miles on CO-15 gets to a right turn onto County Rd. DD/FF. Follow that 1.5 miles to FR-250, bearing left at the fork, then 8.2 miles to an intersection, bearing right to stay on FR-250/255 westbound for 9.2 miles when the road name changes to Alamosa Ave. Continue another 7.1 miles, the name changes back to FR-250 and continues another 1.9 miles to the a fork, stay right to follow FR-380 0.3 miles to the next fork; take that fork left following FR-380 into the campground. Whew! Obviously, this works better with a GPS map but there's no coverage out there!

Boundaries: Upstream access starts west of the formal campground at 37.3743, -106.5812 and continues past the campground to 37.3776, -106.5722. There is private property on both boundaries so please respect fences, etc. To be very clear, this section of the river is <u>unclaimable</u>. There are many claims around here that are outside of the unclaimable area so do not wander around without doing the needed research.

Locale: in the mountains west of Alamosa
Land Type: small river in a mountain valley
Land Manager: USFS

Key Regulations:
• As usual, no gas engines near the campground.

Nearby Attractions & Accommodations:
Stay at Stunner Campground, go fishing or boating at the nearby Platoro Reservoir. Site K-05B in *Finding Gold in Colorado: Prospector's Edition* is on FR-250 downstream of Stunner campground.

Site Number: K-08
Site Name: Alamosa Campground

Alamosa USFS Campground is a very simple single loop in a high mountain valley next to the Alamosa River. It is downstream a bit from the prior site along the same roads.

Local Hints & Cautions:
- Avoid prospecting this area in April and May due to high water conditions.

Gold Finding Tips:
- I haven't explored this spot, but the river has some nice inside bends to explore with your gold pan.

Getting There: From US-285 in downtown Alamosa, it is about 32 miles to the campground, so this site is 12 miles closer to town than the prior one. The campground is on FR-250 at 37.3798, -106.3437; to get there from Alamosa, turn off of US-285 onto CR-10S/CO-370 west bound at 37.4296, -105.8912 and take that 14.1 miles to CO-15 south. Two miles on CO-15 gets to a right turn onto County Road DD/FF. Follow that 1.5 miles to FR-250, bearing left at the fork, then 8.2 miles to an intersection, bearing right to stay on FR-250/255 westbound for 3.2 miles then turn south into the campground. Whew! Obviously, this works better with a GPS map but there's no coverage out there!

Boundaries: Upstream access starts west of the formal campground at 37.3797, -106.3507 and continues to the east edge of the campground at 37.3781, -106.3439. There is private property on the downstream boundary so please respect fences, etc. Like the prior site, this section of the river is <u>unclaimable</u>. There are many claims around here that are outside of the unclaimable area so do not wander around without doing the needed research.

Locale: in the mountains west of Alamosa
Land Type: small river in a mountain valley
Land Manager: USFS

Key Regulations:
- As usual, no gas engines near the campground.

Nearby Attractions & Accommodations:
Stay here perhaps, see comments on the prior site too.

Site Number: K-09A-C
Site Name: The Conejos River

The Conejos River is a small tributary river of the Rio Grande River. It runs from west to east through the mountains along the southern border of Colorado a bit southwest of Alamosa.

There is very little placer gold in the river but a careful panner can find a color here and there. This was enough to lead the original prospectors upriver where they found mother lode deposits in the high country. Today a reservoir is the major feature of that area. Some of the early mines around the town of Plat Oro were started in 1882. For quite a few years there was continued exploration and some mining but no major metals production. Certainly not like there was a bit to the north in Summitville. Panning in this river is challenging and not likely to produce much, although modern prospectors report very small colors and the occasional +30 mesh flake.

Local Hints & Cautions:
- Be respectful of anglers that you are sharing the river with by giving them space and setting up downstream of them rather than upstream.

Gold Finding Tips:
- Smart prospectors will avoid high water in April and May.

Getting There: The following access points are along the river below the old mines, so we know there is gold. To some extent old milling operations, which lost some of the gold they processed, have added gold to the river beyond what the original prospectors found in 1880. From US-285 on the south end of Antonito, head west on CO-17 to the third, then second site below, or continue on CO-17 and turn right onto FR-250 at 37.1330, -106.3505 and continue on this road to the first site below.

Boundaries:
K-09A: Lake Fork USFS CG, gold confirmed, turn off of FR-250 at 37.3094, -106.4764 with boundaries upstream at 37.3108, -106.4780 down to 37.3091, -106.4785 (focus on latitude).

K-09B: Aspen Glade USFS CG; I found gold here myself; turn off of CO-17 at 37.0747, -106.2712 with boundaries upstream at 37.0728, -106,2784 down to 37.0698, -106.2634 which is well past the campground.

K-09C: Ponderosa Campground, a private campground, turn off of CO-17 between mile markers 18 & 19 at 18584 CO-17 (37.1239, -106.3299); must be staying here to dig here of course. 719-376-5857 or www.ponderosaco.com for reservations. Rules: ask permission if you want to use more than just pans; stay on their property - where the nice inside bend is anyway; avoid any anglers; respect the erosion control work. Legal access from 37.123, -106.329 downstream to 37.122, -106.328.

Locale: Conejos River west of the southern San Luis Valley
Land Type: river running through a narrow canyon
Land Manager: USFS

Key Regulations:
- These sites are at USFS campgrounds so no gas-powered equipment without local permission.

Nearby Attractions & Accommodations:
Gorgeous scenery, fishing, multiple camping areas. Stay at the USFS campgrounds along this route or upgrade to a private campground such as Ponderosa Campground, Conejos River Campground, Twin River Cabins and RV Park, or Skyline Lodge. If you choose to upgrade, be sure to ask if they allow guests to prospect in the river.

CHAPTER L: LAKE CITY & THE UPPER GUNNISON RIVER

The Gunnison River was called the Tomichi by the local Utes, a name which is remembered today in the name of one of the creeks which feeds into the river near the town of Gunnison. The river was more recently named for John Gunnison who was killed while mapping a route through the region in 1853. He was killed by Paiute Indians west of here on the Sevier River in what is now Utah.

Prospecting Opportunities

When I wrote the first guidebook, I was frustrated in my efforts to find any placer gold whatsoever, in legal spots around Lake City, despite reports of it in Henson Creek. Over the five years between that book and this one, I worked hard to connect with locals and do further research on places to prospect in or near Lake City. I hope readers enjoy what I was able to find.

The third site below is on another fork of the Gunnison, near the town of Gunnison. My thanks to one of my collaborators from the Facebook group who confirmed access and gold at that site for me just a couple weeks before this book was published!

Site Number: L-01
Site Name: Lake Fork Memorial Park

One of the crown jewels of the Lake City park system, the memorial park sits at the confluence of Henson Creek and the Lake Fork of the Gunnison River. The river is enjoyed by many in summer including lots of kids playing and sometimes anglers.

Local Hints & Cautions:
- You are on town property here so be a perfect guest. Leave no trace, pick up trash, only dig in the river, you get the idea.

Gold Finding Tips:
- This is a very challenging place to pan without much to show for your efforts to be honest, but it is right in Lake City so there's that. At least that was my experience. Digging a little deeper to get into muddy, dense material helps, but given the geology here, it's no surprise that the gold is sparce and small.

Getting There: The Memorial Park is at 254 Spring St., Lake City 81235 with plenty of parking and very easy river access...unless there is a special event going on.

Locale: on the edge of the little town of Lake City
Land Type: smaller sized river running through a valley
Land Manager: Lake City

Boundaries: From Spring St. to the confluence with Henson Creek. If you want to prospect just downstream of the confluence, ask at the River Fork RV Park and Campground, or just stay there of course!

Key Regulations:
- Pans only.

Nearby Attractions & Accommodations:
Stay and maybe dig at the River Fork RV Park and

Campground next door. (112 North Henson St. 970-944-9519 or trailhiker227@gmail.com)

Site Number: L-02
Site Name: Gateview Campground

Gateview BLM campground and the Lake Fork of the Gunnison River around it are unclaimable and a long way from anywhere! Being further downstream from Lake City, the river here has had the opportunity to gather gold from a variety of small and medium hard rock sources. As a result, it is richer than the previous site.

Local Hints & Cautions:
* This location is a 45-minute drive from Lake City. It is very much the middle of nowhere so be prepared to self-rescue from any reasonable problems. I recommend carrying a satellite transponder such as a Garmin InReach or an advanced cell phone.
* This is a narrow gravel road with nowhere to turn around, so RVs are a bad idea.

Gold Finding Tips:
* This area is the upstream edge of part of the reservoir. Since the gold drops when the river hits the still water of the reservoir, and the reservoir's still pool edge moves up and downstream depending on the water level, the possibility of good gold accumulation exists here. This is especially true when the reservoir levels are down a bit and a prospector can get into the bed of the reservoir.
* Try crevicing the bedrock for gold in the cracks.
* I haven't tested this area but have info from another prospector confirming the gold here.

Getting There: From Lake City, head north out of town on CO-149, 20 miles to the left turn onto Blue Mesa Rd/CR-25 (at mile marker 93). Follow that 2.4 miles, then continue straight as the road changes to CR-64/Lake Fork Canyon Rd. after the bridge on the left. (Note: you can also prospect within 50 feet of the bridge; no powered equipment allowed.) Continue 4.7 miles on CR-64 to the campground. This trip takes about 45 minutes to travel 27 miles.

Locale: in the boonies north of Lake City
Land Type: small river in a canyon
Land Manager: BLM

Boundaries: From upstream at 38.3767, -107.2411, downstream to the still water of the reservoir (the actual unclaimable area reaches further north than you will ever be able to go!)

Along the drive north the river is also unclaimable wherever it is within 25 feet of the centerline of the county road, so if you see it that close and it looks worthy of a test pan, try one~

Key Regulations:
• Do not introduce new material into the river. This means no processing of bench deposits above the high-water mark.

Nearby Attractions & Accommodations:
Gateview Campground (has vault toilets and six sites with picnic tables and fire rings, not much else, tent or truck camping is free, no reservations) or several other campgrounds along the way north from Lake City.

Site Number: L-03
Site Name: West Gunnison River Access

This site is just over 1.5 miles west of the city of Gunnison on US-50. It is primarily used by rafters, kayakers and anglers.

Local Hints & Cautions:
• Although this site seems only lightly used, be respectful of other park users. Give any anglers plenty of room if they are already fishing when you arrive.

Gold Finding Tips:

- There is a long inside bend toward the middle of this site. It is probably worth the walk.

Getting There: To access this area from Gunnison on the east, take US-50 west to CR-32 at 38.5217, -106.9956. Head south on CR-32 for ½ mile the first right turn after the bridge over the river. The turn is at 38.5146, -106.9955 onto a dirt drive to a dirt parking lot. Follow the path from there to the river.

Locale: Gunnison River just west of the city of Gunnison
Land Type: river valley
Land Manager: Gunnison Waste Water Treatment

Boundaries: River access from 38.5191, -106.9866 downstream to the bridge over the river that you drove to get here on CR-32.

Key Regulations:
- None noted. Respect any posted rules.

Nearby Attractions & Accommodations:
Mesa Campground, 36128 US-50, Gunnison, 970-641-3186 is just east of here if you want to camp in relative luxury. Curecanti National Recreation area is just to the west on US-50.

Sharing the river with rafters...and my 80+ year old mom!

CHAPTER M: OURAY & THE UNCOMPAHGRE RIVER

A fun fact about the Uncompahgre River - the name was given to the river by the Utes who saw that it naturally ran red every spring during the spring snow melt. The name literally means "red water, spring". The water's color is due to material from Red Mountain washing down into the river, which mostly happens in the spring during the runoff. That's Red Mountain looming in the background in the picture on the facing page.

Prospecting Opportunities

Both of these prospecting sites are the Gunnison River, the first is well upstream of anything in the first guidebook on the North Fork of the Gunnison. You could argue that this site should be in the prior chapter but a glance at a map will quickly illustrate why it is included here.

The second site is barely out of town to the north from Delta. That means it is a bit downstream of Site M-06 and upstream of M-07 in *Finding Gold in Colorado: Prospector's Edition*.

> **Site Number: M-08**
> **Site Name: Hotchkiss Fairgrounds**

Delta County has a nice property in Hotchkiss with the fairgrounds facility and a decent stretch of the North Fork of the Gunnison River. This is good for us because back in 1874, a gold placer mine was discovered and operated upstream of town. Let's see if we can get some of the gold that has moved downstream!

Local Hints & Cautions:
- Respect the private property here.

Gold Finding Tips:
- Explore the inside bend formed along the edge of the fairgrounds.

Getting There: The address is 575 S River Lane, Hotchkiss. The river is on the south end of the property. Take River Lane south to parking at 38.7950, -107.7144.

Boundaries: Upstream access starts at 38.7931, -107.7145 and continues to the Highway 92 bridge at 38.7962, -107.7102.

Locale: Town of Hotchkiss
Land Type: river running across a plain
Land Manager: Delta County

Key Regulations:
- Follow all posted signage.

Nearby Attractions & Accommodations:
Check out the little town of Hotchkiss while you are here.

Site Number: M-09
Site Name: Cummins Gulch

Delta County has a piece of property at Cummins Gulch on the Gunnison River downstream of the town of Delta which includes both banks of the river with fairly easy access. Although the land is classified by the county as recreational, it isn't listed as a park. I wonder if they hope make it into a park someday or put in a formal boat ramp or both?

Local Hints & Cautions:
- Be careful near the railroad tracks. Do not park on the railroad right of way.

Gold Finding Tips:
- The Cobble bar on the east side of the river near the G Road bridge looks interesting but I haven't sampled this site.

Getting There: From US-50/Main St. in Delta, take 5th St. west. It turns into G Road, follow that about 3 miles to the river. Be sure to take the right fork in the road to stay on G Road when it forks away from Sawmill Rd. at 38.7410, -108.1098. To park on the east side of the river, look for a turn off between the railroad tracks and the river. For the west side, turn right onto the little lane just after the bridge at 38.7491, -108.1230 and choose a spot to park that won't block other visitors, including those with boat trailers.

Boundaries: Upstream access starts at 38.7495, -108.1177 and continues to the G Road bridge at 38.7478, -108.1199

Locale: rural area a few miles west of the town of Delta
Land Type: river running through farmlands
Land Manager: Delta County

Key Regulations:
- None noted. Follow any posted rules.

Nearby Attractions & Accommodations:
Check out other prospecting sites in this book and in *Finding Gold in Colorado: Prospector's Edition*. Visit the town of Delta and also the prospecting shop, Mr. Detector, for all your prospecting gear needs.

CHAPTER N: TELLURIDE & THE SAN MIGUEL RIVER

If you would like to see how tough the miners were back in the day and how hard they had it, check out this hike suggested by the Telluride Chamber of Commerce:

Ballard Boarding House Hike

From the far south end of Pine St. in Telluride, at the Bear Creek Trailhead, hike about a mile up the Bear Creek Trail. At that point, in a clearing, there is a log bridge over the creek to the left. Cross the bridge and head up the mountain about 200 feet to find a trail in very good condition that heads upstream (to the south). This trail soon passes an old log structure and continues over a 200-foot-wide avalanche path. Go left at the first fork and then follow the switchbacks up the hillside. The trail swings back across the avalanche path and then into the forest again. Go right at the next fork and continue hiking for a couple hours. At this point, the view opens up displaying a large spire and other rocky formations. Walk up to the middle of the rocky clearing to spot a trail through the rocks. Follow another set of switchbacks up and look to the south to see the remains of the Ballard Boarding House. Take a deep breath and imagine the miners doing this hike every time they came to work. No wonder the mine built a boarding house! Now imagine doing the hard physical labor of underground mining in the late 1800s at this elevation. Those miners had to be pretty tough! On the return hike downhill, it is easy to stay too high and miss the traverse. Beware of steep scrambling if you find yourself off trail. This hike can take up to six hours uphill and about half that to return, so pick a day when the skies will be clear and only do this if you think you are as tough as the miners of the old days! This description is adapted from the telluride.com website which used the book *Telluride Hiking Guide* by Susan Kees as a reference source. It might be wise to get her book in town before heading out on this hike! I am sure Between The Covers at 214 W. Colorado Ave. in downtown Telluride can get you one.

There are no new dig sites in this chapter because I was SO thorough for the first guidebook. If you are interested in prospecting more sites on the San Miguel and on the Dolores River below its confluence with the San Miguel, I have a suggestion. Much of the river is private ranches and mining claims. So, join my Facebook group Finding Gold in Colorado and express your interest. Several active members of the group have claims on this stretch of water and some of them routinely invite others to join them for a dig. I have dug on one of those claims and can tell you, it's a great time, with fun people, in beautiful red rock canyon country! Stand warned, this region is quite remote. Don't expect cell service, groceries, or Starbucks!

Puttin' her back into it!

Oh NO, Mommy, we're
losing gold!

Start 'em young!

Photos on these two pages are
courtesy of Chrissy & Lucas
Wentzel

Can I run the big
one this time?

The Animas River at Baker's Bridge – bedrock prospecting!

CHAPTER O: SAN JUAN RIVER BASIN

This chapter was called "Silverton & the Animas River" in the prior book but that is quite incomplete, so it is changed here.

While this area includes the Animas River, there are also other, nearby waterways which feed into the San Juan River just as the Animas does. These include the La Plata River and the Mancos River. We really should start with the San Juan headwaters first though. The upper San Juan River is in Colorado, running south through Pagosa Springs to the New Mexico border. The other waterways included in this chapter all flow south through Colorado to meet the San Juan in northern New Mexico. So, all of these smaller southwest Colorado rivers are part of the same drainage.

Upper San Juan River
Gold starts getting contributed to the San Juan River in the East Fork of the San Juan to the northeast of Pagosa Springs, so let's start there.

Site Number: O-06
Site Name: East Fork Campground

East Fork Campground is on the east fork of the San Juan River, so the name is not very surprising! This site is easily accessed from US-160 east of Pagosa Springs. The campground is beautiful, well forested with the river at the edge of the campground. You can make a reservation (wise given how easy this is to access, even with a big RV) and there are 26 RV sites as well as room for tent campers.

Local Hints & Cautions:
- The river here is still quite small but entirely unmanaged so during snow melt season it really gets going.
- The hillside down to the river is quite steep in some areas here so watch your step and choose your route carefully.

Gold Finding Tips:
- Check out some of the gentle bends in the river and the big rocks, especially when water levels are low.

Getting There: This site is just an 18-minute drive from downtown Pagosa Springs! Turn east onto FR-667 from US-160 at 37.3795, -106.8992 and drive about a mile to the campground at 373758, -106.8882 where the campground driveway meets the road.

The river runs to the west of the campground but can be accessed both through the campground and from the road. To avoid a day-use charge at the campground, try parking off of USFS-667 just upstream of the campground, in the pull off at 37.3755, -106.8850 and walking down the steep hill to the water.

Boundaries: Upstream access starts north of the campground at 37.3757, -106.8841 and runs SW past the campground to 37.3707, -106.8890 – just keep an eye on the latitude to stay legal as you explore here. Much of this river is claimed so stay in bounds even if there is no noticeable signage.

Locale: forest east of Pagosa Springs
Land Type: small mountain river in the San Juan mountains
Land Manager: USFS

Key Regulations:
- No gas-powered equipment due to the campground.

Nearby Attractions & Accommodations:
Stay at this campground (open 4/15-11/30, 888-785-3234, recreation.gov for reservations) or in Pagosa Springs. Definitely try out Pagosa's Hot Springs "The Springs Resort & Spa". It's one of the best developed hot springs around. Head a little west on US-160 to visit Chimney Rock National Monument to learn more about how the Ancestral Puebloan people lived and studied astronomy.

Site Number: O-07
Site Name: Fireside Cabins

This little resort just east of Pagosa Springs along US-160 offers charming little cabins with indoor plumbing in a gorgeous setting. Obviously, only guests are welcome to prospect here so call them at 970-264-9204 to book a reservation.

Local Hints & Cautions:
- The river here is still quite small but entirely unmanaged so during snow melt season it really gets going.
- The staff was quite amused when I talked with them about gold prospecting so expect to do a little education when you ask about it.

Gold Finding Tips:
- The river here is fairly straight so look for the large rocks that interrupt the flow of the water to create gold concentrations. I have not prospected this site yet.
- Later in the summer after the snow melt is finished, the water levels are low enough to make access very easy and to expose lots of tempting spots to prospect.

Getting There: This site is just a few minutes' drive from downtown Pagosa Springs! Take US-160 east out of town to 1600 E Highway 160, Pagosa Springs.

Boundaries: Upstream access starts at the northeast edge of the campground at 37.2755, -106.9883 and runs SW past the campground to 37.2737, -106.9903, with private property on either side.

Locale: semi-developed area just northeast of Pagosa Springs
Land Type: small mountain river in the San Juan mountains
Land Manager: private property

Key Regulations:
• Ask permission to use powered equipment.

Nearby Attractions & Accommodations:
Stay here obviously! See prior site for other fun ideas.

Site Number: O-08
Site Name: Yamaguchi Park

The San Juan River runs right through town and on the south end of town, it goes past Yamaguchi Park and then the wastewater treatment plant (happily it is downstream of the park, not upstream!). The park itself has ball fields, playgrounds, and toilets, as well as shade and water. It is also only 5.1 miles to the nearest Starbucks in case my friend Wes needs one!

Local Hints & Cautions:
• Don't dig erosion control structures or the narrow spots that have been built to give the rafters a thrill.
• Despite how it looks, the eastern side of the river is private property.
• Do not park in the boat ramp area at the south end of the park if there is any chance you will be in the way of a ramp user.

Gold Finding Tips:
• Just downstream of the artificial narrow spots is a smart place to test.

- When water levels are fairly low there is a nice long cobble bar toward the downstream end of the park.
- If you walk down river into the sanitation district section of the river, there is a nice inside bend to reward your extra efforts.

Getting There: This site is just a few minutes' drive from downtown Pagosa Springs! From US-160, turn south on 8th St., take a left on Apache St. then a right on S 5th St. to the park at 684 S 5th St. There is parking next to 5th St. at the north and south ends of the developed park.

Boundaries: The river runs north to south with access from just south of the condo complex, at 37.2571, -107.0113 downstream to the sanitation district property at 37.2540, -107.0105; if you want to prospect further downstream, I am sure being in the river would be fine to the south boundary of the sanitation district property at 37.2498, -107.0126. Remember, since the river is running north to south, you can just keep an eye on the latitude number as you explore.

Locale: on the south end of the town of Pagosa Springs
Land Type: small mountain river in the San Juan mountains
Land Manager: City of Pagosa Springs

Key Regulations:
- No gas-powered equipment in the park for any purpose.
- No alcohol in the park.

Nearby Attractions & Accommodations:
See the prior sites for suggestions. Pagosa Springs is FUN!

The Upper La Plata River

The gold prospecting history of the La Plata River starts early, especially for this part of the state. Two men, John Moss and C.D. Posten were travelling back east from the California gold fields in 1856. They stopped at the La Plata River and found good gold in their sample pans. In 1873, Moss returned with a group of other prospectors from California. They found gold and also indicator minerals suggesting lode deposits could be upstream. Moss then made a treaty with Chief Ignacio of the Southern Utes which allowed them to mine and farm a 36-square-mile area which encompassed the canyon of the upper La Plata River. By the next summer, Moss had organized supplies and miners to begin operations in earnest. A group from Arizona arrived and also got involved, starting in late 1873. The population grew slowly, no doubt due to the remoteness of this area, but by 1881 the population in the canyon had grown to over 300. As the mines developed, most of the miners lived in large boarding houses. We can see the remains of one on the driving tour.

Upper La Plata Driving Tour

A driving tour up the valley of the upper La Plata River is long and requires 4WD and high clearance at a couple points in the upper section. However, this tour passes several campgrounds, the "almost" ghost town of La Plata City, some old mining ruins and even a couple of operating mines. Near the top of the valley, a large bowl surrounded by Cumberland Mountain and Snowstorm Peak contains many mining artifacts. Bear left and stay on the road headed uphill to the top. At the top you'll arrive at Kennebec Trailhead, above tree line, miles and hours from civilization, with a nice parking lot for about 10 cars, interpretive signage and even cell service!

To find this area, drive US-160 west from Durango to CR-124. The La Plata River is the next drainage to the west of the Animas River. Set your trip odometer to zero as you get off of US-160 at CR-124 to make this a bit easier...about 8 miles northerly up CR-124 is the La Plata City Campground and interpretive site at 37.3874, -108.0771. It is worth a quick stop even though the buildings are gone. As the interpretive signs mention, the town was started based on a minor gold rush and peaked at over 1,000 people. It is hard to imagine that now from the empty field and grove of aspens.

Just a little up the road, several old cabins remain on private property at the site of Upper LaPlata City. Most of them still see occasional use by the owners, having been handed down by one of the original miners, and still in the same family today. The current owner of many of the buildings is the great grandson of an original resident and is a grandfather himself! I was lucky enough to meet him during my first visit to the area. On the east side of the road here, look for the old Assay Office, a red square building with a metal roof and the chimney rising from the center of the roof. It has most recently been used as a summer cabin. On the west side, look for a wooden building very close to the road which was originally the town post office. Boren Creek, just upstream of the old town saw extensive placer mining during the boom. <u>The land along the river is private, or at least the mineral rights are private, so there's nowhere to casually prospect here.</u> Don't worry, we can dig downstream!

Continuing uphill, a very tall fireplace and chimney mark the location of a miners' dormitory for the Gold King mine. The mine itself was just across the creek on the hillside.

At about the 13-mile point, a right turn onto an old 4x4 road for 1.2 miles leads to the Columbus mine. This site includes interpretive signage, a large shaft, and surface equipment. The mine entered production in 1917, producing a telluride ore rich in both gold and silver.

When you get up above tree line to 37.4464, -108.0128, pause to examine the mining ruins including a large double boiler with bizarrely bent smokestacks. This marks the lower end of the Govoner Mine. The upper part has been prepared as a tourist attraction but hasn't been open due to an illness of the owner.

These smokestacks look even bigger in person!

From the parking at the saddle at the top (37.4515, -108.0114), called the Kennebec Trailhead, hiking trails head in both directions. Following the short trail to the right provides some amazing views into the next valley as you climb a small rocky hillside. Rather than continuing downhill at the fork, take the right-hand fork, around the cirque to the Muldoon/Cumberland Mine, a total distance from the parking lot of just a mile. Be very cautious, the trail is narrow, rocky, and unmaintained while perched above a steep hillside. A slip could be deadly! The Cumberland Mine has some of the most amazing views in the state. You'll find the mine portal, a machine shop and the multi-room mining cabin which was used year-round! You'll notice the remains of the rails used to guide ore cars in and out of the mine. The ore was hauled off the mountain by mules as recently as the 1950s. The mining cabin is rare in that it has a front porch, demonstrating the appreciation the miners had for the view. You'll also be amused by the 2-seater outhouse perched on the edge of the cliff. It is barely hanging on so please don't approach it too closely. Be extremely careful exploring this site and please don't remove or damage anything. This mining ruin is quite unique and should be preserved for the next generation of visitors bold enough to come all the way up here!

For those wishing to camp in this area, there are several designated dispersed camping areas along CR-124 including Miner's Cabin at 37.3816, -108.0768, Madden Campground at 37.3874, -108.0771, LaPlata City mentioned above, and Darby Campground at 37.4267, -108.0422.

Mancos & the Mancos River

The little town of Mancos sits along the Mancos River just off of US-160 at CO-184, a bit west of the La Plata mountains. Although it was not started by the gold rush, its history runs deep. The area has been settled since at least the tenth century by Ancient Puebloans. The Old Spanish Trail passed through the area; it was used to connect the Spanish territories of New Mexico and Spanish California from 1829 until the Mexican American war ended in 1848. The town itself was incorporated in 1894 and served as a minor supply center for local agriculture and mining. Today there remain quite a few historic buildings and an excellent bakery/coffee shop.

Site Number: O-09
Site Name: West Mancos River Rec Area

The West Mancos River runs through canyon country northeast of the town of Mancos. On a prospecting tip from a ranger at the Public Lands office in Dolores, I looked into the area. As he suggested, I found reacquired lands we can dig on.

Local Hints & Cautions:
* This area is rather remote, I wouldn't go alone, two cars is wiser.

Gold Finding Tips:
* No clue, I haven't been to this site yet. I'd love to hear about it if you go. Email me at findinggoldincolorado@gmail.com

Getting There: From Mancos, the drive is 10.7 miles and about 20 minutes all the way to the campground, slightly less for the dig site. From Mancos (and US-160) head north on Main St. for a fraction of a mile to the right turn onto CR-42 for 5.3 miles, passing Mancos State Park on the way (an option of a place to stay next to a reservoir). Then continue onto FR-561 for roughly another 1.5 miles. To continue to the campground, continue another 3 miles and turn right onto FR-565 for a quarter mile and find the campground on the left.

NOTE: due to the steep walls of the canyon in the prospecting area, the wise way to walk in is probably to start down on the trail which meets the road at 37.4255, -1082487 (one-mile up FR-561 and just a bit before you get to the prospecting area). This trail goes a few hundred feet, curving downhill, to meet the Jackson Gulch Canal. Then walk upstream along the canal about ½ mile to the downstream end of the prospecting area. The river will be just to your right.

Locale: In the boonies northeast of Mancos

Land Type: smaller sized river running in a narrow valley
Land Manager: USFS

Boundaries: From upstream at 37.4367, -108.2302 downstream for about a half mile to 37.4295, -108.2384. Downstream of that is state property where you would need a lease agreement to dig (the lease money would go to the local, rural school building maintenance fund, by the way).

Key Regulations:
- No gas-powered equipment.

Nearby Attractions & Accommodations:
Further up road 561, onto road 565 for the last ¼ mile is Transfer Campground at 37.4679, -108.2081 (sadly there is a 100-foot cliff between the campground and the river here!). Other housing options include the Double R Ranch at the end point where CR-42 turns into FR-561 or the Mancos State Park.

Site Number: O-10
Site Name: Riverwood RV Resort, Mancos

This can be a fun place to do some panning if you choose to stay here as you explore this part of the state.

Local Hints & Cautions:
- Obviously, you have to be staying at the RV park to dig the river via their property. Ask for current guidelines before digging. Feel free to also ask about using whatever equipment you have in mind.

Gold Finding Tips:
- It is fairly easy to find at least a little gold in your pan here. Sample around to find better concentrations.

Getting There: The Riverwood RV Resort is at 350 E. Grand Ave., Mancos 81328.

Locale: on the edge of the little town of Mancos
Land Type: smaller sized river running through a valley
Land Manager: private property

Boundaries: stay within the bounds of the RV park: behind sites 25, 26, 27, 28 (this is a hint on which campsites you should try to get!). There is only about 75 yards of river on the resort property.

Key Regulations:
- Only prospect here while you are a paying guest. Make your reservations by calling 970-533-9142, via bmancos@gmail.com or at rvparkmancos.com

Nearby Attractions & Accommodations:
The LaPlata driving tour is fairly nearby (see above). Prospecting is also allowed at a city park in town. Details in site 0-06 in my first book.

Site Number: O-11
Site Name: Mancos Library

There is a nice stretch of river behind the library that extends on public ground downstream for a decent distance.

Local Hints & Cautions:
- You are on town and school property here so be a perfect guest. Leave no trace, pick up trash, only dig in the river, you get the idea.

Gold Finding Tips:
- It is fairly easy to find at least a little gold in your pan here. Sample around to find better concentrations.

Getting There: The Mancos Library is at 211 1st St., Mancos 81328.

Locale: right in the little town of Mancos
Land Type: smaller sized river running through a town

Land Manager: Town of Mancos

Boundaries: Use the library parking lot. Walk north to the river to start prospecting at 37.3443, -108.2907 and from there downstream to 37.3440, -108.2914 (just watch your longitude to stay good). This isn't a terribly large area (80 yards of river) but it's a fun spot. Upstream and downstream are private property.

Key Regulations:
• No gas-powered equipment.

Nearby Attractions & Accommodations:
The library of course!

A Driving Tour of the mines just east of Silverton

Just east of Silverton on CR-2, at the intersection with CR-4, at the mouth of Cunningham Gulch sits the remains of the town of Howardville (37.8356, -107.5950). The town was settled in 1874, and Howard's cabin remains on site along with some buildings from the Little Nation mine which was in operation into the 1960s. The town was named for George W. Howard who built the first cabin in 1873, just after the Ute were forced out of the area. The little town was the first county seat of La Plata County with a small courthouse built in 1874. Sadly, that building burned in 1954. When the railroad got as far as Silverton in 1882, Howardville became less important and lost the county seat to its neighbor. The rails did eventually get to town in the 1890s but it was too late. Howardville survived into the 1960s thanks to the Little Nation mine and the Pride of the West mill. You can get a good look at the Pride of the West Mill from the road, most of it is still in place. The remaining town buildings have all been repurposed for modern homes and even an ATV rental operation.

Just up the gulch from here on CR-4 is the Old Hundred Mine which is open for tours. If you stay on the lower road along the creek, you first come to the concrete foundations that remain from the second mill built on this site to serve the Old

Hundred. Look up high on your left to see the remains of the Veta Madre Mine way up there on Galena Peak. About a mile further along is what remains of the Green Mountain Mill just downhill of the road. A big multistory Vertex mine building has been converted into a private home just above the road as it curves to the left – very cool! Further up the gulch are the remains of many other mines, on both hillsides of the gulch. The last mine is the Highland Mary mine up on the right above the trailhead parking. The Highland Mary has a very strange story. It was started by Edward Ennis on the advice of a 'spiritualist' on where to find a "lake of silver". Always short of money, he never properly developed the mine and died insane. Another company bought it for gold production and operated it intermittently until the mill there burned down in 1952. It's definitely worth the drive. When we drove up to the Highland Mary Trailhead, there were three little Subarus sitting in a row in the parking area – that tells you all you need to know about the road conditions!

A little further up CR-2 (past the turn to Cunningham Gulch/CR-4), the turn off into Maggie Gulch is on the right. The road up Maggie Gulch is graded and fairly well maintained but still something for high clearance vehicles ideally. This is another fun drive with amazing views as the road travels above tree line. At the top of the road, a small parking lot sits right next to the Gold Nugget mine. Amazingly, there is much to see here: the old winch, the shaft (with some concrete foundations from the shaft collar fallen in), an ore bin feeding two 5-stamp crushers and a (almost buried in tailings) Wilfley table.

Prospecting in the Animas River

As you will see below, Baker's Bridge is quite infamous. It has also been a fun place to gold pan for decades. I wanted to put it in the prior book but couldn't figure out any legal way to access the river because of all the private land. Some time talking with locals solved the problem.

Site Number: O-12
Site Name: Baker's Bridge

Baker's Bridge is named for the first significant U.S. prospector who came up the Animas River, all the way to Silverton. Unfortunately, he was quite a scoundrel for a couple of reasons. First, he snuck into the area without the permission of the Ute tribe which owned and ruled the area. He did this because he knew he was unwelcome. Second, he built a toll road, including this bridge and went back to eastern Colorado. There he promoted his gold discovery which he said was in the Eureka area just east of Silverton (of course neither town existed at that point). When over 1,000 prospectors rushed into the region, Baker charged each of them to use his toll road and bridge. When they arrived at their destination, they found the gold was small and too sparse to make anyone rich with hand equipment. Within a matter of weeks all of the prospectors headed back...and found themselves paying Baker yet again to use his route. The whole thing was a scam, with Baker making lots of money on road tolls and lots of disappointed prospectors. Some of them were starving and destitute by the time they made it back to civilization. In any case, the road and bridge are now owned by the state. The gold here (and up in Eureka, see my first guidebook for details) is good enough to be fun, as long as you don't have a goal to get rich today!

There are two ways to access the river at this site:
- Hop down the rocks next to the bridge to get to the water. Stay off of the adjacent private property regardless of whether it is fenced or not.
- Enjoy the generosity of the landowner who owns the private property just south of the bridge on the west side of the river. They very kindly allow people to use their land to access the river. This is definitely easier than hopping down the rocks right next to the bridge. Please be a kind visitor, leave no trace and pick up any trash you notice. Obviously respect any signage or fencing you find here - things change.

Local Hints & Cautions:

- The river here is strikingly beautiful, enjoy! When it is running high, this area can be a dramatic show of waterpower, stay out when that is a risk.
- The land along the river is all private property, please respect that and stay in the river.
- Can be crowded with college kids and swimmers during late spring and summer.

Gold Finding Tips:
- It is fairly easy to find at least a little gold in your pan here. Sample around to find better concentrations.
- Test the deposits of sand & gravel sitting on bedrock and large rocks.

Getting There: This site is just east off of CO-550 on CR-250, a reasonably short drive north of Durango or 33 miles south of Silverton. Take CR-250 curving around and downhill to the west side of the bridge over the Animas River, where informal parking is available at 37.4590, -107.7996. Be sure your vehicle is fully off the road.

Locale: La Plata County between Silverton and Durango
Land Type: medium sized river running through a narrow valley
Land Manager: private property

Boundaries: stay in the river, the banks are private property

Key Regulations:
- Stay off of the private property along the river. Hint: It's all private property.
- Leave no trace. Fill holes, pack out trash.

Nearby Attractions & Accommodations:
Be sure to head 33 miles north to Silverton for the awesome museum, mine, and mill tours. You can also prospect at the Eureka site where wily Baker scammed the other prospectors. (See *Finding Gold in Colorado: Prospector's Edition* for lots of information on Silverton.) There is camping available a few hundred yards from this site the next dig site.

Site Number: O-13
Site Name: JW Durango Riverside RV Resort

This resort is pretty fancy and priced to match. It is just up the road and just downstream from the Baker's Bridge site.

Local Hints & Cautions:
- The river here is strikingly beautiful, enjoy! When it is running high, this area can be a dramatic show of waterpower, stay out when that is a risk.
- The land along the river is all private property, please respect that and stay in the river at the campground.
- Obviously, you have to stay at the resort to dig the river via their property...or even to access the river to walk upstream or downstream sampling.

Gold Finding Tips:
- It is fairly easy to find at least a little gold in your pan here. Sample around to find better concentrations.
- Test the deposits of sand & gravel sitting on bedrock and large rocks.

Getting There: This site is just off of CO-550 at 13391 CR-250, Durango CO 81301. The campground is just a couple hundred yards or so south of Baker Bridge itself. This resort is a full-service operation with full hook-up RV, tent, and cabin sites. They offer a playground, heated pool, outdoor movies, an ice cream parlor, café, and convenience store.

Locale: La Plata County between Silverton and Durango
Land Type: medium sized river in a narrow valley
Land Manager: private property

Boundaries: stay in the river, the banks are private property.

Key Regulations:
- Stay off of the private property along the river. Hint: It's all private property.

- Only access the river here while you are a paying guest. Make your reservations by calling 1-970-247-4499 or email info@jwdurango.com
- Accessing the river here is technically against the resort rules for liability reasons so if you choose to go down to the river, you are doing it at your own risk. ONLY go when water levels are low. The resort officially suggests people access the river via the prior dig site instead.

Nearby Attractions & Accommodations:
Be sure to head 33 miles north to Silverton for the awesome museum, mine, and mill tours. You can also prospect at the Eureka site where wily Baker scammed the other prospectors. (See *Finding Gold in Colorado: Prospector's Edition* for lots of information on Silverton.)

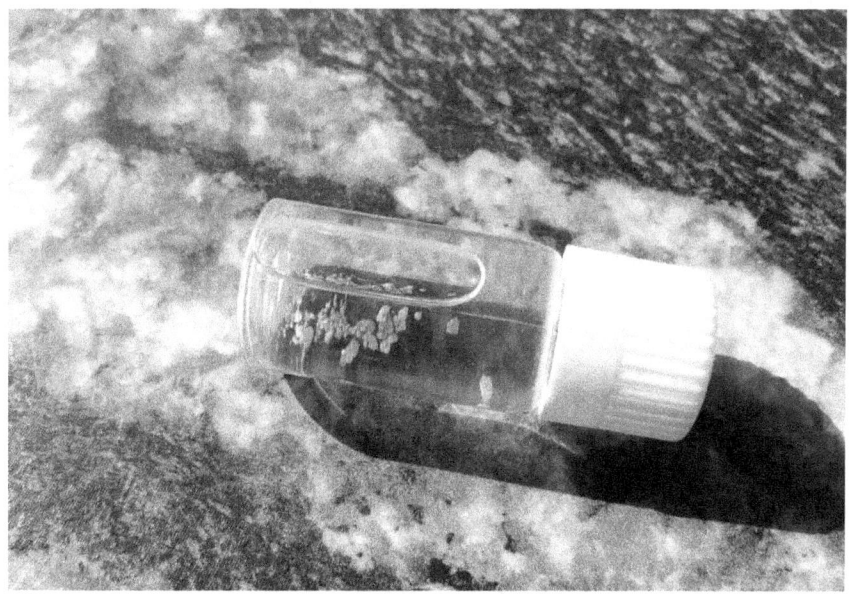
Baker's Bridge gold

A mountain biker whizzes by an ore bin near Rico

CHAPTER P: RICO & THE DOLORES RIVER

The upper Dolores River was explored by the French fur trapper Antoine Robidoux in 1833. He and his group reported finding evidence of Spanish or Mexican miners working the area for precious metals earlier. Robidoux and the other trappers tried their hand at panning in the area but did not find enough gold to bring them back.

Prospecting Opportunity

As I have learned more about how gold bearing waterways interact with reservoirs, I have gone looking for more reservoirs where we are allowed to prospect the point where the river meets the pool of the reservoir. I found one here on the edge of the town of Dolores.

| Site Number: P-07 |
| Site Name: McPhee Reservoir |

The upstream end of this reservoir is on the west edge of the town of Dolores, not too far downstream from site P-05 in the *Prospector's Edition* book. Since the reservoir is used for downstream irrigation, the level of the reservoir moves up and down quite a bit. In addition, the last decade of drought has left the reservoir continuously well below its maximum height for many years.

Local Hints & Cautions:
- Access to paydirt here requires some walking to bring a wagon or pack light. I recommend a wagon since the trail is flat, even paved, for much of the way.
- This site can be explored at any time of year but there will generally be more access from late summer through the winter into spring – before the spring melt gets going in earnest.

Gold Finding Tips:
- The special situation here is that gold drops at the point where the fast-moving river hits the still water pool of the reservoir. Since that intersection moves upstream and downstream over time, there are multiple deposits in the park. So, what you are looking for is a cobble bar or bed in the general area upstream of the current reservoir pool.

Getting There: The parking lot at 37.4710, -108.5179 is at the west end of Central Ave. which is one block north of CO-145/Railroad Ave. From there, follow the old dirt road further west into the park. Look for interesting areas in the river to the south (left) of the old road. Explore around!

Locale: Dolores
Land Type: Sometimes-flooded valley, sometimes wide with a river running through it.
Land Manager: US Bureau of Reclamation

Boundaries: None that are relevant.

Key Regulations:
Do not introduce new material into the basin of the reservoir.

Nearby Attractions & Accommodations:
Enjoy the little town of Dolores. Try dig sites right in town from the prior guidebook *Finding Gold in Colorado: Prospector's Edition.*

Your smile could be
here next!

CHAPTER Q: WESTERN COLORADO RIVER COUNTRY

Traveler's Note: If you are traveling east of New Castle on I-70, along the Colorado River, be sure to also take a look at Chapter S. If your plans continue east on I-70 from Dotsero toward Vail, also take a look at Chapter T.

Prospecting Opportunities

The sites in this chapter are sprinkled over about 110 miles of I-70 in western Colorado and Utah, with the last site just over the border into Utah because I just couldn't resist! The amounts of gold to be found at these sites vary more than might be expected so prospecting along this route offers some surprises. There is a fair amount of overlap between this chapter and Chapter Q in *Finding Gold in Colorado: Prospector's Edition* in the sense that many of these new sites fit in between the ones in the prior book.

Site Number: Q-10
Site Name: BLM Rec Site

Just west of New Castle and east of Silt, sits a large BLM recreation area on the south side of the river.

Local Hints & Cautions:
- Although this site seems only lightly used, be respectful of other park users. Give any anglers plenty of room if they are already fishing when you arrive.

Gold Finding Tips:
- This is a long inside bend so sample close to shore.
- There is an island at the midpoint of this site which would be fun to explore with a gold pan but access is challenging.

Getting There: To access this area from New Castle on the east, get off I-70 at Exit 105, head south across the river to turn right onto CR-336 aka Colorado River Road. (Heading north from the interchange is the route to Site Q-01.)

To access this area from the west (Silt), get off I-70 at Exit 97, head south, turning left onto the River Frontage Rd. Follow it to CR-218 which quickly turns into CR-311 and crosses the river. You'll drive right by site Q-02 on the island. Once off the island, continue south to a T-intersection where you are forced to turn. At the T-intersection, take a left to continue on CR-311 headed east. The road meanders a bit and turns into CR-336.
If this seems complicated, just get off at the Silt Exit 97, put the GPS coordinates listed below into your Google Maps app and follow the prompts...or just drive to New Castle and back track a little as described above.

At the east end of this site, there is parking along county road at 39.5488, -107.5851 with walk-in access to the river from there. There are also a couple of pull-off parking areas on the north side of the county road where you have the option of parking and walking north to the river, although I am unsure of why you'd do that with a bunch of prospecting gear in hand!

The better parking choice for most people is the onsite parking access from CR-335 at 39.5459, -107.5957, then drive almost all the way to the river where there is a nice dirt parking area. You can also follow the road headed east from the parking area, parallel to the river for about 300 yards or so, to another smaller parking area at 39.5483, -107.5922. This spot puts you very close to a nice inside bend. Just follow the little trail north to the river and explore along the bend.

Note: this site is just upstream of site Q-02 (Silt) in the first guidebook.

Boundaries: River access from 39.5491, -107.5835 to 39.5470, -107.5974 for a little over ¾ mile of river.

Locale: Colorado River valley between Silt and New Castle
Land Type: river valley
Land Manager: BLM

Key Regulations:
• None noted.

Nearby Attractions & Accommodations:
Other dig sites near here in the original book or in Chapter S. Lodging options in New Castle or Silt.

Site Number: Q-11
Site Name: Silt KOA Campground

The KOA campground in Silt, formally named Colorado River KOA Holiday is indeed right on the river and walking distance from downtown Silt at the same time.

Local Hints & Cautions:
• Ask for a site that backs up to the river so you can pan while you enjoy your campsite!

Gold Finding Tips:
• There is a lot to explore here. Access is enhanced by a bridge to an island just behind the office. There is also access between campsites (don't cut through other people's campsites!) at 39.5426, -107.6601 where there is a simple boat ramp. Obviously, don't block the ramp or dig into it.

Getting There: To access this area, get off I-70 at Exit 97, head south, turning right onto the River Frontage Rd. Follow it to the office at 629 River Frontage Rd.

Note: this site is just upstream of the next site but on the other side of the river.

Locale: Colorado River valley right in the town of Silt
Land Type: river valley
Land Manager: Private property

Boundaries: River access from 39.5421, -107.5531 to 39.5425, -107.6627 for a little over 1/2 mile of river.

Key Regulations:
• You must be a guest of the campground to prospect here.
• Follow campground rules, if in doubt, just ask, they are very welcoming!

Nearby Attractions & Accommodations:
Other dig sites near here in the original book or in Chapter S.

Site Number: Q-12A, B
Site Name: Silt River Preserve

The Silt River Preserve is being improved and enhanced over time so it may be different by the time you visit. The local government intends to add picnic areas and a variety of other amenities; there are already toilets and a parking area.

Local Hints & Cautions:
• Be very respectful of other users as you share the site. Stay away from the fishing pier if there are anglers using it, for example.

Gold Finding Tips:
• The usual routine for the Colorado River, look for larger cobbles and gravel.

- If water levels are low enough for safety, feel free to explore the small islands just north of the river bank as they are part of the unclaimable area here.

Getting There: From I-70, Exit 67 for Silt. Then:
Q-12A: South Side: Head to the south side of the interchange and then east on River Frontage Rd. Take the first right (after the Holiday Inn Express) onto 16th St./CR-311. Follow that over the river (past site Q-02 from the first book), to the first right turn onto Dry Hollow Rd., continuing straight onto Rifle-Silt Rd. to the last right turn onto the park access road at 39.5342, -107.6627. Follow the access road north, past Highwater Farm to the actual park.

Q-12B: North Side: Head to the south side of the interchange but turn west on River Frontage Road and follow it to the end of the road.

Locale: Rural Mesa County just south of Silt
Land Type: wide valley with the Colorado River
Land Manager: Town of Silt

Boundaries:
Q-12A: Upstream: 39.5387, -107.6586. to downstream at 39.3418, -107.6685. Since this area runs east to west, you can just focus on staying within the longitude numbers.

Q-12B: 39.5426, -107.6681 down to the bridge.

Key Regulations:
- Follow all posted rules.
- Do not attempt to access the large island at the upstream end of the park. There is a bald eagle nest on the island, so all use is forbidden.
- On the north side, keep gas powered equipment 50 feet from the bridge. Panning and sluicing closer is fine.

Nearby Attractions & Accommodations:
The little town of Silt actually has a lot to offer in the way of food and lodging.

Site Number: Q-13
Site Name: De Beque BLM Rec Site

De Beque (or Debeque on some maps) was named for a Dr. De Beque who settled here in 1884, just a few years after the Utes were forced to move elsewhere. It has been a ranching town from the start with boom-and-bust cycles driven by the oil and gas industry.

Just east of De Beque on the south side of the river, there's access to the river for about 1/3 of a mile along the leading half of a long inside bend.

Local Hints & Cautions:
* Bring your own water, there's no drinking water available (or any other facilities) and this site gets hot!

Gold Finding Tips:
* The usual routine for the Colorado River, look for larger cobbles and gravel.
* If water levels are low enough for safety, feel free to explore the islands just north of the riverbank as they are part of the unclaimable area here. Just be sure to stay between the longitude limits in the boundary info below.

Getting There: From I-70, Exit 62 for De Beque. Head to the south side of the highway and go easterly on the I-70 Frontage Road which becomes 46 ½ Rd. all the way to 39.3501, -108.1728 where you will find parking off the road. This is also the beginning of the prospecting area. There is an irrigation ditch between the road and the river. If the ditch is full of water, you may choose to backtrack a couple tenths of a mile to the point where there is a bridge over the ditch. Then just follow the informal road on the other side of the ditch back upstream a wide pull-out area at 39.3501, -108.1634 (due north of the other parking area and right across the ditch).

Locale: Rural Mesa County, east of De Beque
Land Type: dry forest along a braided river

Land Manager: BLM

Boundaries: Upstream: 39.3505, -108.1681 to downstream at 39.3510, -108.1728. Since this area runs east to west, you can just focus on staying within the longitude numbers.

Key Regulations:
- Respect any signage and leave ranch gates as you find them (either open or closed).

Nearby Attractions & Accommodations:
There isn't a lot to the town, but you can find the basics here. This is an area where it is possible to see wild horses and burros so keep an eye out for them or even ask about their whereabouts while you are in town. The locals often know!

Site Number: Q-14
Site Name: De Beque River Park

The town maintains a river park with a boat ramp just off the expressway. This is the sort of prospecting site you could use as a quick break from driving; it's that close to the I-70 exit! Amenities include shade trees, picnic tables and toilets.

Local Hints & Cautions:
- Bring your own water, there's no drinking water available (or any other facilities) and this site gets hot!

Gold Finding Tips:
- The usual routine for the Colorado River, look for larger cobbles and gravel.
- Local advice says to look for black clay/sand for the best gold. Check the gravel bar under the Roan Creek Rd. bridge.

Getting There: From I-70, Exit 62 for De Beque. Head north toward the river and town on Roan Creek Rd. (it's the only choice!) and turn off the road to the left or right just before the bridge over the Colorado River. Your total drive from the bottom of the offramp to the parking is just 1/3 mile!

Locale: Rural Mesa County, east of De Beque
Land Type: river in a dry valley
Land Manager: Town of De Beque

Boundaries: Upstream The I-70 bridge to downstream at 39.3284, -108.2146. Since this area runs east to west, you can just focus on staying within the longitude numbers. NOTE: once upstream of the parking area on the east side of the Roan Creek Rd. bridge over the river, it is important to stay in the riverbed because much of the dry land is private property.

Key Regulations:
• Pans and sluices only.
• No digging vegetation of any sort.
• Avoid the boat ramp.
• Fill all holes, even in the water.

Nearby Attractions & Accommodations:
There isn't a lot to the town, but you can find the basics here. This is an area where it is possible to see wild horses and burros so keep an eye out for them or even ask about their whereabouts while you are in town.

Site Number: Q-15
Site Name: Plateau Creek Confluence

This is a quirky spot. It's an outside bend on the Colorado River and the parking is on the Creek but we are definitely interested in the river. So why stop here with a gold pan? Well, the flow of the creek into the river creates a disturbance in the river flow which drops gold just downstream of the confluence. It's not common to see a cobble bar on an outside bend but you will here!

Local Hints & Cautions:

- Bring your own water, there's no drinking water available (or any other facilities) and this site gets hot!
- Only park in the pullout area where it is safe and legal.

Gold Finding Tips:
- Focus on the large cobble bar just downstream of the confluence.
- I haven't sampled the creek, but it could have gold too, who knows?!

Getting There: From I-70, Exit 49 CO-65. Drive about ¼ mile to parking pullouts on both sides of the state highway. From there walk to the creek and follow it downstream, under the bridges to the river.

Locale: Rural Mesa County, between De Beque and Palisade
Land Type: semi-arid lands along a creek and river
Land Manager: BLM & CDOT

Boundaries: None relevant.

Key Regulations:
- Do not park on the shoulder of the interstate or the Colorado state highway.
- No gas-powered equipment within 50 feet of a CDOT structure.

Nearby Attractions & Accommodations:
Just 5 miles west is the exit for Palisade which has lots of things going on, places to eat and stay.

The next site in sequence is Site Q-04 in the *Prospector's Edition*.

Site Number: Q-16   **Site Name: Old Bridge/Herkie's Launch**

Just downstream off of Exit 44, for Palisade, is our next opportunity to dig as we go downstream. The old highway bridge here is now called North River Road. Just past that to the south is Herkies Launch where local boaters put in and take out their boats. I sampled this spot in the fall when the water was low. I noticed an island at the upstream end of this area but didn't get to test it, if you do, let me know how it is!

Local Hints & Cautions:
- If you park at Herkie's Launch, think carefully about where you park so you aren't blocking boaters with trailers from being able to maneuver.
- Only park in the pullout areas where it is safe and legal.

Gold Finding Tips:
- The inside bend is just upstream of the N River Rd. bridge.
- Feel free to also explore the island upstream of the inside bend if you can safely get to it.
- Walking downstream from the launch parking also includes some interesting spots, especially during low water periods.

Getting There: From I-70, Exit 44 onto US-6 toward Palisade. Drive about 1/3 mile to parking on the river side of the road at 39.1210, -108.3205. You will find a large dirt lot there with parking both to the left and right once you pull off the highway.

Locale: Mesa County, between De Beque and Palisade
Land Type: semi-arid lands along a creek and river; some tree cover along the river
Land Manager: BLM & CDOT

Boundaries: None relevant.

Key Regulations:
- Do not park on the shoulder of the interstate or the Colorado state highway.
- No gas-powered equipment within 50 feet of a CDOT structure, although it is difficult to get that close here anyway.

Nearby Attractions & Accommodations:
The town of Palisade is just a couple miles down US-6.

The next site downstream is Q-05 in the *Prospector's Edition*.

> **Site Number: Q-17**
> **Site Name: Clifton Nature Park**

Although the parking and trails for this site are part of the Mesa County Open Space parks, the river itself is managed by the BLM. This means you are crossing through the park but not actually prospecting there.

Local Hints & Cautions:
• Bring a wagon if you want to carry very much. It is about a ¼ mile walk from the parking area to the river.

Gold Finding Tips:
• Once you get to the river, explore upstream along the inside bend.
• If the water is low enough to get out to the islands, explore the upstream end of the island that has a lot of permanent plants on it.

Getting There: From I-70, Exit 37, hop onto US-6 westbound (toward Clifton/Grand Junction). Drive 3.4 miles (including two left turns at 32 Rd. and D Rd.) to park on the south side of the road at 39.0627, -108.4522. You will find a large dirt lot there.

Locale: Mesa County, between Grand Junction and Palisade
Land Type: semi-arid lands along a braided river; minimal shade
Land Manager: Mesa County Open Space for access, BLM for the river from the high-water mark down

Boundaries: If you stay on the river side of the recreational path, you are good.

Key Regulations:
• Follow posted rules.

Nearby Attractions & Accommodations:
The towns of Palisade and Grand Junction. **The next site downstream is Q-06 in the Prospector's Guide**. It is virtually across the street to the west...in fact you can take the rec path downstream from Q-18 to Q-06.

Site Number: Q-18A, B
Site Name: Eagle Rim Park/Las Colonias Park

On the south side the river, Eagle Rim Park sprawls across much of an inside bend in south central Grand Junction. The north rim of the park is undeveloped and inviting to the local prospector. Just downstream on the north side of the river, Las Colonias River Park offers an area of braided river with islands.

Local Hints & Cautions:
- This site is part of a developed park. Be respectful of other users.
- There is a pedestrian bridge over the river here. It is wise to walk out on the bridge to survey the river. This will let you see current conditions and where you may want to start sampling.
- There is a bit of walking on good paths here to get from the parking to the digging so a wagon may be helpful.

Gold Finding Tips:

- There are multiple areas on the south side of the river, both upstream of the bridge and downstream, where cobble bars form.
- The north side is more of a mystery with multiple threads of river depending on the season. I have not prospected that side myself.

Getting There: Both sites are just off of Riverside Parkway. From I-70 Exit 31, head south on Horizon Drive to 26 ½ Rd., which becomes 7ᵗʰ St. Finally, turn left at Riverside Parkway.

Q-18A: Eagle Rim Park. The actual park is on the south side of the river but for prospectors, it is wiser to access this site from the north side. This leads to less walking, fewer contacts with non-prospectors, and a good chance to view the river from the bridge before choosing a site to dig. With that in mind, instead of just using Google Maps to go to the park, set the app for directions to go to Las Colonias Park and boat ramp instead. From Riverside Parkway, travel 0.9 miles east to the right turn onto Winters Ave. Follow that to its end at the boat ramp and pedestrian bridge parking at 39.0545493, -108.5460655. Avoid parking in a way that will interfere with vehicles pulling trailers.

Q-18B: Las Colonias River Park. From the turn onto Riverside Parkway, travel ½ mile to a right turn onto Riverfront Dr. and then the park is the first turn on the right. While you could park in the same spot as Q-15A, most of Q-15B is downstream of that parking so it can be more efficient to park at the Las Colinas Park's Dog Park parking lot at 39.0555, -108.5514; this lot also has much more space for vehicles and no concerns about boat trailers.

Locale: Grand Junction
Land Type: semi-arid lands along a braided river with minimal shade
Land Manager: City of Grand Junction

Boundaries:
Q-18A: from as far upstream as you'd want to go (just stay along the river since there is a residential area above the river) to 39.0523, -108.5489 with private property beyond there.

Q-18B: from the pedestrian bridge, downstream to the US-50 bridge.

Key Regulations:
• Follow posted rules.

Nearby Attractions & Accommodations:
All of Grand Junction. The next site is just downstream as well.

Site Number: Q-19
Site Name: Dos Rios Park

This park is so new that in 2023 as I was confirming GPS coordinates and driving details, Google Maps still didn't show it in their satellite imagery. The park, as you might guess from the name is at the confluence of two rivers: the Colorado and the Gunnison. This is exciting for us because the Gunnison carries a lot more gold than the Colorado at this point so once we are downstream of the confluence, the likelihood of good gold in our pans goes up quite a bit.

Local Hints & Cautions:
• You'll definitely want hip waders if you are going to explore much of this area due to the braided nature of the river (multiple islands and shallow water areas intermixed).

Gold Finding Tips:
• It'll take a fair amount of exploring to get a sense of this area. Exploration is MUCH easier when water levels are low. I visited during high water, so I wasn't able to do much.
• Due to the way the two rivers meet, it is important to get out onto the big island to get to its outside bend because that is where the Gunnison River water is flowing. This

means working your way across the first narrow channel directly in front of the parking area and playground. Remember Safety First!

- Focus on areas with cobbles. Ignore the many areas here that are silty and lack larger rounded rocks.

Getting There: From I-70, Exit 31, get onto Horizon Dr. southbound. Turn left onto 261/2 Rd., then right onto Patterson Rd., then left onto 25 Rd. which curves gently left to become Riverside Pkwy. Finally, turn right onto Dos Rios Dr. to the left onto Dos Rios Ct. and to parking on either side of the road at 901 Dos Rios Court. You will find a series of diagonal parking spots there. The destination is 39.0594, -108.5714.

Locale: Grand Junction
Land Type: semi-arid lands along a braided river with minimal shade
Land Manager: City of Grand Junction

Boundaries: If you stay on the river side of the recreational path, you are good from the railroad bridge, downstream to Hale Ave. (39.0623, -108.5774).

Key Regulations:
- Follow posted rules.

Nearby Attractions & Accommodations:
All of Grand Junction. The next site downstream is Q-08 in the Prospector's Guide.

Site Number: Q-20A, B
Site Name: Town of Whitewater

Whitewater is a little town south of Grand Junction on the Gunnison River, so this site is a bit of a side trip away from the Colorado River. We know the Gunnison River carries good gold so it's no surprise there is gold in Whitewater. There isn't a lot of access in this little town due to all the private property but there are a couple spots.

Local Hints & Cautions:
• Be very respectful of private property adjacent to these access sites.

Gold Finding Tips:
• While there are solid reports of gold here, I have not sampled this area myself.

Getting There: From I-70, Exit 37, follow 32 Rd. which is also CO-141 south. South of town, take the left turn onto US-50 which also continues to be CO-141. Take US-50 2.5 miles to the right turn to continue on CO-141. From here follow the instructions to each site below. This entire drive only takes about 15 minutes.

Locale: Southern Mesa County in the town of Whitewater
Land Type: semi-arid lands along a braided river with minimal shade
Land Manager: Mesa County and the Union Pacific Railroad

Boundaries:
Site Q-20A: Whitewater Boat Ramp. The boat ramp is at 38.9717, -108.4541, reached from the intersection of US-50 and CO-141 as follows. Go southwest on 141 for a few hundred feet and take the left onto Desert Rd. Follow that 0.8 miles through a left turn and then a right turn to Mill Tailing Rd. Take the hard right onto Mill Tailing Rd. and follow it 0.6 miles to the parking area near the river. Be very careful of the railroad as the crossing to get to the river is unregulated. Never park or loiter near the tracks. The upstream prospecting boundary is at 38.9690, -108.4556 while the downstream boundary is close to the north edge of the parking area at 38.9725, -108.4542.

Site Q-20B: CO-141 Bridge. From the US-50/CO-141 intersection, follow CO-141 about 0.4 miles, to the parking area on the side of the highway just over the Gunnison River bridge at 38.9837, -108.4520. Prospecting access is from upstream of the bridge at 38.9816, -108.4526 to downstream of the bridge at 38.9877, -108.4531 (given the north/south orientation of this site, you can just watch your latitude number as you walk along). Stay close to the river at this site since the area away from the river and highway is all private.

Key Regulations:
- Follow posted rules at the boat ramp. Stay away from the railroad tracks except to cross them when safe.
- No power equipment use allowed within 50 feet of the CO-141 bridge.

Nearby Attractions & Accommodations:
All of Grand Junction and little Whitewater too.

Site Number: Q-21
Site Name: Westwater Boat Launch

While this site is just across the border in Utah, I've included it because it's close to Colorado, has decent gold (for a fascinating reason), and it is a take-out point for boaters who float through the National Conservation Area between Fruita and here.

The gold is good because, while the current river shape here is an outside bend, just over 100 years ago the river was flowing ¼ mile to the east and this area was part of a large inside bend that had formed over many centuries.

Utah State Flag

Local Hints & Cautions:
- The road is well used by boaters but is rarely maintained so be prepared to get unlucky and find that those last 9 miles from the interstate to the boat ramp are a slower drive than you'd expect - it usually is about a 20-minute drive, but conditions can vary.
- There are pit toilets in the parking area. Definitely bring your own water, especially in hotter weather.

Gold Finding Tips:
- Ignore the fact that this "looks" like an outside bend and just sample anywhere away from the boat ramp and boat traffic if any.
- Because this whole area was previously an inside riverbend, dry washing may also be productive here (gold was reported in a bench deposit of river gravels 40' above the river near here).

Getting There: From I-70, take Exit 227, turn south toward Westwater 0.4 miles, then left onto Old Highway 6 &50 for just 0.1 miles. Continue straight onto BLM192/Harley Dome Road for 5.2 miles, then left onto BLM 191/Down Harley Dome/Westwater Ranch Rd. 3.6 miles (don't miss the left turn right after the railroad tracks) to the boat launch at 39.0888, -109.1024. There you will find two parking lots. The first is on the left before you get to the river, the second is where the road dead ends alongside the river.

Locale: Southeastern Utah
Land Type: semi-arid lands along a braided river, minimal shade
Land Manager: BLM

Boundaries: From upstream at 39.0903, -109.1024 (across from the island which is upstream of the first ramp parking) downstream to the boat ramp itself (but don't block boat traffic). If you want to dry wash, I suggest exploring the area north and west of the first parking lot, between the road and the railroad right of way. Please fill any holes to avoid BLM concerns about erosion.

Key Regulations:
- Follow any posted rules.
- Pans only due to threatened and endangered aquatic species, no sluices, dredges, etc.

The Yampa River in Winter

CHAPTER R: CRAIG AND THE YAMPA RIVER BASIN

I wrote about this area in Volume One, so let's just dive into the new dig sites!

Site Number: R-09A, B
Site Name: Steamboat Lake State Park

This first site in this chapter was one that I wanted to include in the prior book, but the whole area was a construction project at the time due to renovations being done on the dam. I am excited to include it now because this is a very historic area.

As the rangers will tell you, Steamboat Lake Reservoir was built on the largest placer mine in the area. Today only the fish can get to the gold in the reservoir, but we have a chance nearby as long as we play really, really nicely. Being a state park, there is a fee to access the area unless you have a pass. I encourage people to pay for your pass at the discounted rate when you register your vehicle.

Local Hints & Cautions:
• This park is also an active fishing zone. Do your best to avoid scaring the fish where someone is fishing.

Gold Finding Tips:
• There are multiple streams which flow into the reservoir. All are accessible to explore but can be a challenge in that these areas in the park can be boggy with a lot of sediment.

Look for dense gravels.

- There are nice little bends in the flow out of the dam which is called Ways Gulch. Even in a little creek, those bends matter.

Getting There: The state park visitor center is at 61105 County Road 129, Clark, 80428. From US-40 north of Steamboat Springs, turn right onto CR-129 and follow that for just over 24 miles to the park.

R-09A: Deep Creek, Willow Creek, and Dutch Creek all flow into the reservoir from the north. From Sunrise Vista CG, Dutch creek is just west, and Willow Creek is just east. Deep Creek is easier to reach by hiking northwest from the parking lot at the visitor center or crossing CR-129 and following the dirt road that heads due north along the outside edge of the park. Grab a park map when you arrive to see what I mean.

R-09B: Turn left off of CR-129 at Sage Flats Rd. (40.8048, -106.9426) and follow it to the end at 40.7963, -106.9472. There is a parking lot, boat ramp and vault toilets. Then walk over to the outflow of the dam, just to the south.

Boundaries:
R-09A: The upstream edge of Deep Creek in the park is at 40.8157, -106.9505 well uphill from CR-129; Willow Creek's upstream limit is at CR-62; Dutch Creek's upper limit is at 40.8123, -106.9784, upstream to the northwest of CR-62.

R-09B: Ways Gulch: Upstream access starts at the outflow of the dam and continues to the edge of the state park property at about 40.7871, -106.9410. This is the area most likely to be productive since the water flow is consistently high enough to move the lighter material along.

Locale: Northern rural Routt County, north of Steamboat
Land Type: state park in a gorgeous setting
Land Manager: Colorado State Parks

Key Regulations:

- The rules in a state park are quite stern, nothing but pans and shovels, no removal of concentrates, put everything back as you found it. Technically, you can't even take home a pinch of gold in a snuffer bottle since everything in a state park belongs to the state, but (as they say on the state parks website) they sort of look the other way if it fits in your pocket. This is only true if you show respect for the park and its resources by cleaning up after yourself and not annoying other park users. You've been warned.
- If you find anything significant (like a nugget!), the rangers at the visitor center would like to see it. If you are lucky, they will put it on display in the visitor center, and you will be famous!

Nearby Attractions & Accommodations:
Camp on site or stay in the fun town of Steamboat to the south. Be sure to check out the little museum and shops in Hahn's Peak Village just across the county road from the park.

Paul and Kevin review the Museum of NW Colorado display about the Hahn's Peak mining district

---

Site Number: R-10A-C
Site Name: Little Snake River

The Little Snake River weaves in and out of Colorado and then heads south to meet the Yampa River in Dinosaur National Monument. It is in far northwest Colorado where there are far fewer people than pronghorns or turkeys! This particular spot is unclaimable because it has been set aside for a reservoir, which so far has not been built.

Local Hints & Cautions:
- This site is in the literal middle of nowhere. Be sure to tell someone where you are going and assume there will be no help wandering by if you need it.
- Some of the access points include steep drops to the river. Use good judgement, avoid unsafe situations!

Gold Finding Tips:
- There are great inside bends at these sites, along with some braided channels to explore.
- Away from the river, look for bench deposits and dry washes to sample. There may be some decent dry washing opportunities in this area.
- I have not explored this area myself yet. Good luck and be safe.

Getting There: From the town of Maybell on US-40, turn right onto CO-318 at the west edge of town and drive 14.5 miles northerly to CR-21. Then follow one of these directions.

R-10A: Turn right and follow CR-21 for 14.7 miles, to the left turn onto CR-26 which leads to the river in about a mile. From the informal parking at the end of the road, follow the trail north into the prospecting area.

R-10B: Turn right and follow CR-21 for 10.1 miles to a left turn onto a dirt track at 40.6957, -108.1988, then follow the dirt track west toward the river for about 3-4 miles to a fork in the road. Take the right fork to a very informal parking area at

40.7003, -108.2115. From there walk west to the river.

R-10C: Turn left and follow CR-21S for 4.7 miles to CR-10. Take the sharp right onto CR-10 and follow it for 3 miles to the prospecting area. This area includes a bridge over the river and multiple parking spots near the river on both sides. Look for dry washes and inside bends that provide safe access down to the river.

Boundaries:
R-10A: The upstream edge is 40.7674, -108.1823 just above the beginning of the inside bend. Access continues to the center of the bend at 40.7649, -108.1849 after which point, the mineral rights are privately owned.

R-10B: River access from just upstream of the little creek at 40.7012, -108.2123 downstream to 40.6997, -108.2133; all of the river here is within 5-600 feet of the parking area.

R-10C: Both sides of the river are accessible from latitude 40.5508 south to 40.5437; access to the southern portion is easier from the west side of the river along CR-153.

Locale: Northwestern Moffat County, northwest of Maybell
Land Type: arid high desert
Land Manager: BLM

Key Regulations:
• No restrictions identified.

Nearby Attractions & Accommodations:
RV and tent camping as well as gas and a convenience store in Maybell. Full Services upstream on the Little Snake River in Baggs WY. Dinosaur National Monument with cool dinosaurs, camping, rafting, and hiking.

The Upper Colorado River has its own charm.

CHAPTER S: UPPER COLORADO RIVER BASIN

The Colorado River begins its journey in Grand County on the west side of Rocky Mountain National Park. The first gold strikes in the upper waters were in Bowen Gulch, north of Grand Lake on the edge of the park (which was created years later of course). The first formal mine was established in 1879 when James Bourn and Alexander Campbell started the Wolverine Mine. However, they couldn't make it work and the area was abandoned by 1885. That said, the upper Colorado River gets gold added to it from multiple sources. Willow Creek west of Granby and the Blue River near Kremmling are two noteworthy contributors and both worthy of our attention as prospectors. (The Blue River is included in Chapter E.) There are also other rivers like the Eagle (Chapter T) and smaller creeks which contribute at least some gold. The result is varying amounts of gold along the drainage, most of it quite fine in nature.

Willow Creek
Northwest of Granby in Grand County, wraps around Gravel Mountain and then flows into the Colorado River. For those interested in geology, Gravel Mountain is a fun puzzle. The current theory is that the gold was left behind as Gravel Mountain eroded downward over the millennia. However, no hard rock deposits that would serve as the source of the gold have been found anywhere on the mountain or elsewhere in the Willow Creek drainage.

Gold discoveries here were reported in the Denver Rocky Mountain News on June 17, 1871. Several modest scale placer mines were operated with about 40 men working the area as of 1880. Placer gold production continued intermittently through the late 1940s but never amounted to an awful lot – in the peak year of 1934 just 11.19 ounces of gold were produced across 9 operations. The largest attempts at commercial operations were at Denver Creek and Gold Run Creek. Much of the creek is reportedly unworked virgin gravels.

In any case, Gravel Mountain IS gold bearing. There are active claims in quite a few areas here on parts of Willow Creek, Kaufmann Creek, and various other small creeks running off of the mountain. The good news for us is that when CO-125 was built through this area, a federal law (Public Land Order 3806) was passed banning new claims within a 200-foot strip centered on the road. So, as of 1965 new claims were forbidden near the road and indeed there are no pre-1965 claims as of now. As you will see below, this gives us a large area to explore and good gold to be had as long as we follow the rules and stay away from any active claims beyond the exclusion zone.

This general area is about a half hour northwest of Granby by way of US-40 and then CO-125. This area can get crowded on the weekends but is absolutely empty during the week.

Nearby Granby has grocery stores, a craft brewery, restaurants, and a variety of other services. One fun business to check out is "Showboat's Drive-By Pie". You drive down an alley, past the side of a house(?!) and order pies (and other food such as chicken soup, burritos, lasagna, chicken pot pie, etc.) to go. Very quirky, a bit expensive, but delicious! (www.showboatsdrivebypie.com or 970-887-1111) There is also cell service in town, which you won't find while out in the gold fields here. Granby is also quite close to the western access road into Rocky Mountain National Park.

Site Number: S-01A-L
Site Name: Willow Creek along CO-125

As mentioned in the chapter introduction, a large stretch of Willow Creek is 'mostly' unclaimable and definitely gold bearing. When the federal funding for this road was provided in the 1960s, a federal law made it illegal to file a mining claim within a 200-foot-wide stretch centered on the roadway. However, this stretch of creek is only "mostly" unclaimable since any bits of the creek, which are far enough from the highway, are still claimable...and should be avoided unless you have checked for claims.

Local Hints & Cautions:
- Water levels in Willow Creek vary seasonally but are unlikely to be an issue at any time except during peak runoff (usually in May).
- Mosquitoes are a challenge here so bring some bug spray.
- The country rock here is mostly sedimentary with some granitic intrusions. The sedimentary rock tends to be flat and "slabby". This makes setting up a sluice easier, but classifying is harder since the flat rocks will tend to cover your classifying screen.

Gold Finding Tips:
- The gold here ranges in size from the usual small stuff up to some decent sized flakes and small chunky bits just barely big enough to pick up. Of course, your results may vary - I hope you find <u>even</u> <u>bigger</u> stuff than I did!
- The gold here is usually, but not always, in the regular spots. Start by looking for bigger rocks on, or just downstream of, an inside bend. I found that most times, digging deeper led to less gold instead of more but your results may vary. Sample pan regularly to confirm that you are still on the gold.
- Densely packed material with a bit of clay content mixed into the gravel and cobble seemed to pay the best (as is often the case!).

Getting There: From Granby, head west on US-40 2 or 3 miles to CO-125. Turn northerly onto CO-125 and drive north about 9 miles until you get to CR-21/FR-112. This is the downstream edge of the public access and is marked by a large USFS monument style sign on the east side of the road.

Locale: rural Grand County NW of Granby
Land Type: medium sized creek running through mountain valley
Land Manager: USFS

Boundaries: The upstream edge of this long stretch of river is at 40.3005, -106.0705. The downstream edge is marked by a large USFS monument sign at 40.2137, -106.0489.

Parking spots along this route can be used as dry camping sites if you don't mind being next to the highway. They include:
S-01A: 40.2967, -106.0747
S-01B: 40.2920, -106.0788
S-01C: 40.2899, -106.0802
Trail Creek Trailhead (see below)
S-01D: 40.2789, -106.0859
S-01E: 40.2723, -106.0849
S-01F: 40.2600, -106.0765
Denver Creek Campground (See below)
S-01G: Road 108.1, cross bridge to park on west side of creek at 40.2467, -106.0729
S-01H: 40.2417, -106.0682
S-01I: 40.2332, -106.0619
S-01J: 40.2309, -106.0589
S-01K: 40.2246, -106.0566 (lots of beaver ponds, tough to prospect)
S-01L: 40.2144, -106.0507 at FR-112. Park on either side of the bridge. There is access to bedrock just below the bridge. Crevicing here produced gold. There is also good dry camping just up FR-112 on the far side of the bridge from the highway.

Key Regulations:
- Non-powered equipment only unless you are on a valid mining claim with appropriate paperwork (NOI, POO) due to the fragile nature of this high-altitude riparian environment.
- Do not dig near bridges, erosion control structures or other improvements.
- **Stay within 100 feet of the highway centerline at all times.** Further from the road you are most likely on someone's claim. (Don't count on claims being marked, claim markers have a way of disappearing for various reasons.)

Nearby Attractions & Accommodations:
There is plenty of USFS and dispersed camping available here. Quite a few of the dispersed camping sites are just off the shoulder of the road and are listed above as parking sites for prospecting. If you show up during the week, there will most likely be lots of spots to park and to camp. On summer weekends it is a whole different story as all the obvious free camping spots all fill up. There are also lots of ATV and hiking trails - the local USFS ranger station in Granby can provide maps. The western access into Rocky Mountain National Park is about an hour east of here.

Site Number: S-02
Site Name: Trail Creek Trailhead

This is a designated recreational site on Willow Creek so even creek areas more than 100 feet from the highway centerline are open to public access. There are also about 6 dispersed camping sites here and a bridge over the river.

Local Hints & Cautions:
See the first site for this chapter.

Gold Finding Tips:
- See the first site for this chapter.
- Notice the way the slope of the creek gets gentler here. Could be a good place for gold to accumulate right? There is

also a big bend in the creek with a large cobble bar just downstream of the dispersed camping area.

Getting There: Trail Creek Trailhead is on the east side of CO-125 at 40.282251, -106.084518 and has signs on the highway as you approach from either direction.

Locale: rural Grand County
Land Type: medium sized creek running through a narrow mountain valley
Land Manager: USFS

Boundaries: Upstream: 40.2807, -106.0859 to downstream at 40.2825 -106.835 (Note: This site is essentially north to south so use the latitude to guide your location.)

Key Regulations:
　　See the first site in this chapter.

Nearby Attractions & Accommodations:
See the first site in this chapter.

Site Number: S-03
Site Name: Denver Creek Campground

Denver Creek Campground sits on both sides of the highway. The uphill side has the advantage of almost no mosquitoes, less road noise, and great views down valley. The downhill side has the advantage of being on the creek. This is a designated recreational site on Willow Creek so even creek areas more than 100 feet from the highway are open to public access within the boundaries below.

Local Hints & Cautions:
* This campground is only open seasonally. Confirm with the Sulfur Ranger Station if you plan to camp before Memorial Day weekend or in late season.

Gold Finding Tips:
- See the first site in this chapter.
- There are a number of interesting creek bed features in the campground such as islands, bends and bedrock on the far shore.
- You can also pan in little Denver Creek which passes through the campground on its way downhill to meet Willow Creek.

Getting There: On CO-125 at 40.253942, -106.078195 with a driveway on each side of the highway.

Locale: rural Grand County
Land Type: medium sized creek running through a high valley
Land Manager: USFS

Boundaries: On Willow Creek from upstream at 40.2644, -106.0794 to 40.2608, -106.0767 regardless of how far from the highway you are. Also, on Denver Creek from Willow Creek upstream to the east edge at -106.0770.

Key Regulations:
- Only dig within the high-water mark of the creek.
- Do not disturb growing plants.
- Do thorough reclamation after digging – remove dams, push tailings into holes, etc. Make it hard to tell you were there!

Nearby Attractions & Accommodations:
See the first site in this chapter. Additionally, there are quite a few campsites at this campground including some pull-through sites in the east/upper loop. The pull-through sites are not very level so bring ALL your leveling blocks if you RV!

Site Number: S-04
Site Name: Willow Creek Reservoir Campground

This campground sits on just downstream of the reservoir. It is our last chance to prospect Willow Creek before it travels through private property to its Colorado River confluence.

Local Hints & Cautions:
- This campground is only open seasonally. Confirm plans in advance with the Sulfur Ranger District USFS Station.

Gold Finding Tips:
- Don't let the fact that we are just below a reservoir discourage you. The gold has been accumulating here for thousands of years and was never commercially exploited.
- What is more likely to discourage you is that the waters here are marshy and boggy most of the year. Finding paydirt can be a real challenge.

Getting There: At the west edge of Granby, turn off of US-40 onto US-34. Head north and turn left onto the road to the reservoir at 40.1403, -105.8944 traveling westerly about 3.2 miles to the dam and campground.

Locale: rural Grand County
Land Type: medium sized creek just below a dam
Land Manager: USFS

Boundaries: Upstream 50 feet below the dam, downstream along the campground to 40.1425, -105.9349

Key Regulations:
- Only dig within the high-water mark of the creek.
- No public access to the waterway from 6/01-8/31.
- Do not disturb anything placed by dam construction, etc. If someone spent money on it, you should stay away.

Nearby Attractions & Accommodations: On-site camping! Near Granby and not so far from Rocky Mountain National Park.

Site Number: S-05
Site Name: Hot Sulphur Springs

Hot Sulphur Springs has been a resort town for many decades. The hot springs resort is a fun place to relax and even to watch a train roll by from a hot springs pool. I've heard rumors that hot springs guests occasionally moon the Amtrak trains! The cute little town stretches along the river and has maintained a town park on the river since the early days of the town's creation. The park provides river access, amenities and even camping sites for a reasonable price.

Local Hints & Cautions:
• Please be friendly with any locals you happen to meet at the park. They are likely to be surprised to meet a gold panner here!
• The water can be fast and furious here during spring runoff, be careful or better yet, come when the water is low since gold access will be much better.

Gold Finding Tips:
• Think about where slow water and eddies would be during spring runoff, especially just downstream from the bridge into the park.
• Be prepared for very, very small gold and not too much of it. You may find none at all in many pans, so tenacity is key.

Getting There: From US-40 in the town of Hot Sulfur Springs, follow Spring Road through town and over the bridge into Pioneer Park at 40.074, -106.107. You will quickly see a day use parking area on your left.

Boundaries: Upstream of the developed park from 40.083, -106.097 downstream to 40.072, -106.1120. Since this is a one mile long, east to west stretch of river, just watch your longitude as you get close to the edges of the town property.

Locale: on the edge of the town of Hot Sulphur Springs
Land Type: medium sized river along a partially developed city park
Land Manager: Town of Hot Sulphur Springs

Key Regulations:
• Only dig within the high-water mark of the river.
• Do not disturb live plants or dig sideways into the bank.

- No firearms or motorized equipment allowed in the park.

Nearby Attractions & Accommodations: On-site camping (although you may be woken by trains at night)! The Hot Sulphur Springs resort. Near Granby and not so far from here to Rocky Mountain National Park.

A side adventure - S-06: Walton Creek

If you are headed west from Hot Sulfur Springs, past Kremmling, on US-40 toward Steamboat Springs, there are rumors of gold in Walton Creek up near Rabbit Ears Pass near the boundary of Grand County and Routt County. The easiest legal access is at Walton Creek Campground, just off of US-40 at 40.3819, -106.6874. The creek runs along the north edge of the campground and is unclaimable both there and where it is within 300 feet of the centerline of the highway (this is due to Public Land Order 3069, issued in 1963). I have not tested this area yet. Walton Creek eventually meets the Yampa River in Steamboat Springs. It might also be possible to sample the creek down in town somewhere but there is a lot of private property to avoid. Stop in at city hall to ask about park access.

Site Number: S-07A-C **Site Name: Colorado River – Trough Road**

Upper Colorado River sites between Kremmling and State Bridge are best accessed via Trough Road which travels west from CO-9 just south of the town of Kremmling and south of the Colorado River. (I suppose a boat would be even better since this area is popular with river runners and anglers.) As we head west on Trough Road, the first available dig site is actually on the Blue River, so it is listed in Chapter E of this book, where it is the last site in the chapter (since it is the furthest downstream).

Local Hints & Cautions:

- This section of the river can get crazy-busy during peak rafting season. At other times, before and after the high-water season, it is basically abandoned.
- I have highlighted areas with good access and plentiful parking. There is gold all along the river here so if you find a spot with public access (no private property issues) and cobbles, give it a try!

Gold Finding Tips:
- Go for a walk along the river looking for cobble deposits. I found gold in every place where I tested around cobbles, usually multiple colors per pan.

Getting There: From CO-9, just south of Kremmling, take Trough Rd. westerly along the south side of the river to each access site. After crossing into Eagle County, the road crosses the river to follow along on the north side. This road is also Grand County Road 1, Colorado River Headwaters Byway, and Eagle County Road 11. Most people just call it the Trough Road. The west end of this stretch of road meets CO-131 at State Bridge. From here heading south on CO-131 will connect with I-70.

Locale: rural Grand and Eagle Counties
Land Type: river running through narrow mountain canyons and wider valleys
Land Manager: BLM, except as noted

Boundaries: Irrelevant, this whole area is unclaimable under Public Land Order 74366 – Upper Colorado River Special Management Area.

Key Regulations:
- Follow posted regulations including the BLM day use fee. ($5 as of 2020)
- Don't park in a trailer parking site in the various parking lots unless there are no other options.

Nearby Attractions & Accommodations:
Dispersed camping areas (such as the BLM dispersed camping

site just a mile east of site S-07A at about mile marker 9.5, 39.9907, -106.4918) and near other river access points along Trough Road.

S-07A: Pumphouse Recreation Area (39.9874, -106.5092) is very popular as a put-in point for rafters floating the upper Colorado River. This site provides plenty of parking, bathrooms, water, and camping sites. From CO-9 take Trough Rd. about 10.5 miles to CO-106 north to the river (1.3 miles).
- Heading upstream from the day use parking area, the Gore Canyon Trail provides a pretty walk and many access points to the river.
- Heading downstream, the Argentine Trail runs from the end of the campground for 4.4 miles to Radium.

S-07B: Radium Area (39.9516, -106.5561) – From CO-9, take Trough Rd. 15 miles, turn right onto CR-11, then 2.4 miles to the river (and bridge). Geologists report multiple gold bearing gravel terraces including at river level, 20' up and 100' up. Larger colors in bedrock cracks. All of this is slowly washing down into the river over long periods of time. In the river, where we have access, gold can be found both upstream and downstream of the Radium bridge.
- On the south side of the river, head upstream following the fishing trail to access sites, decent color about ¼ mile upstream of Sheephorn Creek (the creek alongside CR-11 as you drive north toward the river) which reaches the river at 39.9547, -106.5503. Look for exposed bedrock below the highwater mark and check any pockets of gravel.
- Hike the upstream trail to Radium Hot Springs and sample around that area, then have a soak in the springs. Access the trail across the road from Mugrage Campground, which is at 39.9513, -106.5420, then walk about 2/3 mile north to the river. The actual hot springs are at 39.9510, -106.5407.
- Sample likely looking areas just downstream of the bridge on the south side where the river takes a nice inside bend around the rafter parking area and Radium River

Access Campground (which is just before you cross the Radium Bridge).
- Overall, this area has some of the better gold in the upper Colorado, but you may have to walk a bit and sample to find it!

S-07C: Rancho del Rio CG (private) 39.8941, -106.6073, 4199 Trough Rd, Bond 80423, 970-653-4431, ranchodelrio.com, small day use fee. I included this site more for the amenities than the gold. You can get lunch here for example. Much of the river here is fine sediments that are unlikely to be gold bearing. After some searching, I found cobble/gravel deposits to sample and found gold on the upstream end of the property.

Site Number: S-08A, B
Site Name: Colorado River at State Bridge

State Bridge looks as historic as it is. This spot is very popular with the local DIY rafting community which can lend a party atmosphere to the place at times.

Local Hints & Cautions:
• This section of the river can get crazy-busy during peak rafting season. At other times, before and after the high-water season, it is basically abandoned.

Gold Finding Tips:
• When the water is low, the large cobble bar just 100 feet downstream of the boat ramp is a good spot to start. I found gold everywhere I tested around cobbles, usually multiple colors per pan.

Locale: rural Eagle County
Land Type: river running past a bridge at a bend in the river
Land Manager: BLM

Boundaries: Irrelevant, this whole area is unclaimable under Public Land Order 74366 – Upper Colorado River Special Management Area.

Getting There:

S-08A: pull off just upstream of State Bridge on the North side at 39.8566, -106.6431 parking in the Day Use area. Hike the little informal path 50' right down to the river. (on left end of parking area).

S-08B: On the south side of river between the boat ramp and highway bridge on a large cobble bar. Pull off at the boat ramp parking area at 39.8575, -106.6484, DO NOT dig at the actual boat ramp. This area has bathrooms and picnic tables but no drinking water. A beautiful setting, with a camp store just across the river for cold drinks, etc. Best access in summer when water levels have dropped. Day parking is at 39.8576, -106.6482; if you want to explore more, hike the dirt road upstream from the parking area.

The Colorado River along the Colorado River Road:

Sites downstream of State Bridge/McCoy and the confluence of the Colorado River and the Eagle River are best accessed via CO-131 to the Colorado River Road. Travel North on CO-131 from I-70 to State Bridge or take the Trough Road from US-9 just south of Kremmling (as described above) to get here. The Colorado River road is a left turn at 39.9228, -106.7372 as you travel northerly after crossing State Bridge and just past the town of McCoy.

While there are many legal access spots on the river along CO-131 north of State Bridge, I don't recommend them due to the steep cliffs in the area with the exception of the site below. IF you choose to brave it, keep in mind that the legal access is interrupted by a couple of ranches and ends just before the town of Bond at 39.8697, -106.6896 at private property.

Site Number: S-09
Site Name: Two Bridges River Access

A patch of BLM managed, state-owned, land.

Getting There: Turn off of CO-131 at 39.8900, -106.7017; be sure to avoid blocking the boat ramp area or leaving a mess.

Boundaries: Upstream of the bridges there is an inside bend and continued access for over ½ mile. Downstream of the bridges is private land, do not enter.

Site Number: S-10A-J
Site Name: Colorado River Road

These sites are on a stretch of well-maintained, dirt road which eventually transitions to pavement as it heads downstream. It travels from the little town of McCoy to Dotsero where the Colorado River has its confluence with the Eagle River (see Chapter T for the Eagle River information). At Dotsero, the Colorado River Road meets I-70. For much of the route, it travels alongside a well-used railroad track through a river valley few people see unless they are on a train or floating the river. Along this route, the river is unclaimable and much of it is open access. However, be sure to respect private property. Just keep an eye out for fences and signage.

Local Hints & Cautions:
- The railroad which runs along the river here is active with both freight traffic and Amtrak. Be especially careful to park well away from the railroad's right of way and to avoid being near the rails when a train goes by. If you moon a train (yes, it's a tradition if you are in the river when it goes by), I'm going to say I told you not to!!
- Feel free to try other spots along the river road. There are patches of private land mixed in along the way so don't jump any fences!

Getting there: To find the Colorado river road, head a bit north of State Bridge on CO-131 to a left turn at 39.9228, -106.7372.

S-10A: Catamount Bridge Rec Site, turn off CR-30 (Colorado River Road) at 39.8907, -106.8319 just after crossing to the south side of the river. Avoid the boat ramp, the inside bend is between the bridge and the boat ramp.

S-10B: Derby Junction. This spot has pit toilets and river access on an inside bend. Pull off of the river road at 39.8426, -106.9406 to park.

S-10C: Pinball Boat Access & rec site. Avoid the boat ramp at this site and be sure to park where you will not hamper rafters with trailers. I usually see vehicles parked on the shoulder of the road rather than along the driveway to the ramp. Find your spot to park near 39.8336, -106.9469. NO prospecting is allowed at the rec site, so you will have to walk just a little to get upstream of 39.838, -106.945 or downstream of 39.8370, -106.945 (you can just focus on the latitude number on your GPS since the river is running north to south here). Check out the bend and island downstream of the rec area.

S-10D: Informal pull off at a big inside bend at 39.8097, -106.9612 (need I say more?).

S-10E: Red Dirt Creek. Pull off of the river road onto Red Dirt Creek Road at 39.8043, -106.9719. Park as close to the bridges as practical and hike upstream under the bridges to the inside bend.

Side Note: S-10F: Red Dirt Creek Road. Turn off on the road as described above but then follow it about to a number of dispersed camping sites. Please stay on established routes and campsites. The parts of this road near the river are Eagle County Open Space so camping, fishing and prospecting are all good.

S-10G: Cottonwood Island Rec Site. Pull off at 39.7126, 107.0446 where you will find a boat ramp (again!) and pit toilets. Try the big gravel bar upstream of the boat ramp based on local advice.

S-10H: Lyon's Gulch Rec Site. An informal dirt pull-off at 39.7008, -107.0668 (with an inside bend) or more correctly, just a stone's throw downstream at 39.6690, -107.0729. At the second site there is a boat ramp to avoid, informal camping and pit toilets.

S-10I: Informal pullout at an inside bend, at 39.6810, -107.0734 ...hooray for inside bends!

Side Note: at 39.6687, -107.0678 you will find the turn off to CR-600, Deep Creek Road which becomes Coffee Pot Road further up. This decent dirt road leads to 6 dispersed camping sites with picnic tables and fire rings (check current fire rules before having a fire here of course).

S-10J: Large informal pull-out at 39.6536, -107.0645 on a large inside bend. This is the last big bend before the Colorado River Road meets the I-70 Frontage Road at a roundabout. While this side of the river is public land, the opposite side of the river is private, so don't cross over. Public access is from 39.658, -107.072 to 39.651, -107.064 which includes all of the big inside bend and quite a bit more!

Locale: rural Eagle County
Land Type: river running through an arid valley
Land Manager: BLM

The Colorado River along I-70/US-6:

The last sites for this chapter are along I-70, and US-6, which parallels I-70 and sometimes is the same road, from about the confluence with the Eagle River (Chapter T) downstream toward the towns of Silt and Rifle where chapter Q begins (see the *Prospector's Edition* for details).

Site Number: S-11
Site Name: Dotsero Landing

This site is just downstream of the prior site but is on the opposite side of the river. The main attraction here is the inside bend under the bridges and the islands in the river. There is plenty of parking, picnic tables, some shade, and toilets.

Local Hints & Cautions:
• As usual along the Colorado, avoid the boat ramp and stay out of the way of boaters with trailers.

Gold Finding Tips:
• Try the inside bend just downstream or the islands if they are accessible.

Getting There: From I-70, Exit 133 for Dotsero, follow the frontage road on the north side of the freeway Cotton Lane at 39.6495, -107.0609. Dotsero Landing will be immediately on the left.

Locale: Eagle County adjacent to I-70 and near the town of Dotsero
Land Type: river running through a wide mountain valley
Land Manager: BLM, CDOT

Boundaries: Upstream: 39.651, -107.063 to downstream at 389.648, -107.061; either side of that is private property on this side of the river.

Key Regulations:
Follow regulations posted on site. Day use only.

Nearby Attractions & Accommodations:
The town of Dotsero & I-70. Just upstream to the east is the Eagle River which is covered in Chapter T.

Site Number: S-12
Site Name: Two Rivers Park

This site is just downstream of the prior site but is on the opposite side of the river. The main attractions here are the inside bends, but this site does include some walking on rec paths to get to the good spots.

Local Hints & Cautions:
- While the river is federally managed public lands, the lakes are owned by a local authority, the Two Rivers Metro District. Do not attempt to prospect in the lakes.

Gold Finding Tips:
- Try the inside bends, a bit upstream, or downstream, of where the rec path reaches the river.
- The upstream cobble bar is right across the river from the confluence of the Eagle and Colorado Rivers.

Getting There: From I-70, Exit 133 for Dotsero, US-6 to CO River Rd, S. from roundabout to bus stop roundabout at 39.6461, -107.0640. From here follow the sidewalk south to the river. Then follow the rec path either upstream or downstream to inside bends with cobble bars.

Locale: Eagle County adjacent to I-70 and on the south edge of the town of Dotsero
Land Type: river running through a wide mountain valley
Land Manager: BLM

Boundaries: about ¾ mile in either direction from the point where the sidewalk reaches the river at 39.643, -107.063.

Key Regulations:
   Follow regulations posted on site. Day use only.

Site Number: S-13A-E
Site Name: Glenwood Canyon (I-70 Rest Areas, etc.)

From the Two Rivers Park site, the Colorado River runs west into Glenwood Canyon. There are several Colorado Department of Transportation rest areas that afford access to the river.

Local Hints & Cautions:
- Remember the main purposes of a highway rest area do not include providing a place for you to park. Please park well away from the rest area facilities, even if this means you

will have to walk a bit. This way typical rest area users have easy access to the bathrooms and picnic area you aren't using.

- Some of these rest areas are also used by boaters or anglers. Be respectful of their use: stay out of the way of boaters trying to launch or land their craft. Do not approach an angler without first catching their eye silently and getting permission to come closer. Otherwise, they will be concerned about you spooking the fish.
- Avoid high water. Some parts of the canyon are narrow and can create dangerous water conditions even when the river seems perfectly fine elsewhere. High water? Move on and live to pan another day.

Gold Finding Tips:
- All the usual tricks seem to work here. Look for more natural areas and dig on cobble bars or IN the river near shore where the water is quieter.

Getting There: From I-70, Exit 133 for Dotsero, head west on I-70 to each of these sites in sequence. You will also see a rec path for pedestrians and bicycles along part of the canyon. This can also be used to gain access to the river to sample areas further from the sites designated below.

Locale: Glenwood Canyon between Dotsero and Glenwood Springs
Land Type: river running through a narrow, steep canyon
Land Manager: CDOT

Boundaries: Not relevant in most cases, except at noted. Stay off of any apparent private property of course.

Key Regulations:
   Follow regulations posted on site.

S-13A: Bair Ranch Rest Area (Exit 129), Restrooms, picnic area, river access.

S-13B: Hanging Lake Parking Area (exit 125).

S-13C: Grizzly Creek Rest Area (Exit 121).

S-13D: No Name Rest Area (Exit 119), walk south to river. Note: this site is adjacent to private property. Please stay on the rest area property.

S-13E: Glenwood Canyons Resort & No Name Bar & Grill (Exit 119), This is private property adjacent to the No Name Rest Area. They welcome prospectors who are staying at the resort. Get a riverfront site if you want easy access to pan: Sites 18 and 19 are the best. Avoid the boat ramp area.

Site Number: S-14A-C
Site Name: Glenwood springs

Much of both banks of the Colorado River through the town of Glenwood Springs are owned by the city or CDOT so we have lots of access here.

Local Hints & Cautions:
- Remember this is city property and people will be watching you, so play nice and clean up after yourself. Fill those visible holes and smooth out any tailings piles.
- Avoid high water. Weather events just upstream in the canyon can create dangerous water conditions even when the river seems perfectly fine elsewhere. High water? Move on and live to pan another day.

Gold Finding Tips:
- All the usual tricks seem to work here. Look for more natural areas and dig on cobble bars or IN the river near shore where the water is quieter.

Getting There: From I-70, Exit 116 for Glenwood Springs, head upstream or downstream to get to each of these sites. Details below.

Locale: City of Glenwood Springs

Land Type: river running through a narrow valley
Land Manager: Glenwood Springs and CDOT

Boundaries: Not relevant in most cases, except at noted. Stay off of any apparent private property of course.

Key Regulations:
- No removal or damage to plants is allowed.
- Follow city park regulations posted on site.

S-14A: Two Rivers Park. From Exit 116 go north to W. 6th St, west to 740 Devereaux Rd., then south to the large parking lot.

 This is the largest formal park in town and is at the confluence of the Roaring Fork and the Colorado Rivers. There are a lot of facilities here including restrooms, picnic areas and a playground. I confirmed gold downstream of the boat ramp. Think about high water deposition and you'll find more of the fine gold I saw in my pan. Access is from the upstream edge of the park at N River Dr., downstream to the bridge (Devereux Road).

S-14B: The City of Glenwood Springs also owns all of the river from between Devereux Rd. and the river on the south side of the river all the way around the giant inside bend to the Midland Ave. bridge. Check out the exposed bedrock near the local soda bottling plant. This area also includes the town's whitewater kayaking park. Avoid any conflicts with boaters and obviously avoid tampering with anything the city has built in the river to facilitate kayaking. Parking can be very limited here during warm weather due to all the kayakers, avoid this area when busy.

S-14C: Then back on the north side, the city owns the long

inside bend from the Midland Bridge downstream past all the offices off of Riverine Rd. and Gilstrap Ct., with a rec path along the river through this section. Walk the path to any promising looking areas.

Nearby Attractions:
There are many tourist attractions and lodging options in Glenwood Springs. If you are inclined to head south from here toward Aspen (where accommodations are much more expensive), flip to Chapter W for the Roaring Fork Valley sites.

Site Number: S-15
Site Name: South Canyon Boat Ramp

This boat ramp is just off the freeway and provides ample parking...especially when it isn't boating season. Lucky for us, the best prospecting is when the water is low, while the best boating is during high water.

Local Hints & Cautions:
- Avoid this area during high water season. Prospecting won't be productive and the kayakers will be swarming the area.
- The south bank of the river is also open here but be careful of the railroad if you venture across the bridge to the south side.

Gold Finding Tips:
- All the usual tricks seem to work here. Look for more natural areas and dig on cobble bars or IN the river near shore where the water is quieter.
- Look for areas that are eddies (near shore where the water circles back upriver. Prospect the turbulent boundary (between the regular flow and the back upstream flow) to find a paystreak.

Getting There: From I-70, Exit 111 for Glenwood Springs, head to the south side of the highway and downstream to get to the parking area.

Locale: Colorado River canyon 5 miles west of Glenwood Springs
Land Type: river running through a narrow valley
Land Manager: Garfield County

Boundaries: Not relevant, take a walk if it appeals to you!

Key Regulations:
- Do not block the boat ramp or dig there. Give right of way to boaters, especially if they need access to an eddy where you are working.

Site Number: S-16
Site Name: I-70 Eastbound Parking Pull-off

On the eastbound side of I-70 between Exit 105 and Exit 111, there is a large, paved pull-off parking area on the river. This is used by truckers, the boating community and travelers who just need a quick break. This site is not accessible from the westbound lanes at all.

Local Hints & Cautions:
- Be respectful of parking area users with big rigs and trailers. Do not block their ability to maneuver through the lot smoothly.

Gold Finding Tips:
- This site is one big inside bend.

Getting There: From I-70 eastbound (only!) past exit 105, look for the pull-off at 39.5724, -107.4782.

Locale: Colorado River valley between Glenwood Springs and New Castle
Land Type: river running through a widening valley
Land Manager: CDOT

Boundaries: Not relevant, take a walk if it appeals to you.

However, the inside bend is the interesting part and that is what the parking area is built on top of!

Key Regulations:
- Do not disturb anything CDOT spent money on. This especially means the anti-erosion rocks placed to protect the freeway and parking area.
- No gas-powered equipment within fifty feet of CDOT structures.

> **Site Number: S-17A-B**
> **Site Name: New Castle Town Parks**

The town of New Castle started with one house in 1883, not long after the Ute Indians were removed by force in 1881. It grew due to the discovery of high-quality coal deposits with easy access to them. The coal was needed for the smelters used in hard rock mining. By 1908 the coal mines were closing down, and the new excitement was the gold found north of town in East Elk Creek (see site Q-01 in my prior book) but the gold up there proved less significant than would be needed for a long-term mining industry. Since then, the town has seen oil shale boom and bust but really survives on farming, ranching and some tourism including hunting.

Local Hints & Cautions:
- Be respectful of other park users. Give any anglers plenty of room if they are already fishing when you arrive.

Getting There:
S-17A: Grand River Park at 633 Riverview Dr. Go east from the bottom of the I-70 Exit 105 ramp on CR-335 to the large parking lot at 39.5647, -107.5110, think about where the river would drop gold during high water. Boundaries are upstream from 39.5637, -107.5077 around the bend to 39.5645, -107.5143, you can just focus on the longitude to make sure you are in bounds.

S-17B: Coal Ridge Park. Go west from the bottom of the I-70

Exit 105 ramp on the south side of the river to 6705 CR-335 (39.5678, -107.5248). Twelve acres with a boat ramp, bathrooms, picnic tables, etc. Plentiful parking. Panning further downstream from the boat ramp may be smart. Boundaries are from the bridge over the river, downstream past the park's parking lot to 30.5698, -107.5306. There is a rec path that runs west from the parking lot to the downstream boundary and beyond. Please don't prospect past the downstream boundary since the rec path is on a right of way over private lands.

Locale: Colorado River valley in the town of New Castle
Land Type: river running through a widening valley
Land Manager: Town of New Castle

Boundaries: Not relevant, take a walk if it appeals to you. However, the inside bend is the interesting part and that is what the parking area is built on top of!

Key Regulations:
- Do not disturb anything the town spent money on. This especially means the anti-erosion rocks placed to protect the park from erosion.

Nearby attractions: There are a couple of locally run hotels in town or you can camp at Elk Creek Campground, 581 CR-241, 970-984-2240. There are a half dozen or so restaurants; support small town Colorado!

> The rest of the Colorado River sites are in Chapter Q, with some overlap. My apology for any confusion this causes.

CHAPTER T: EAGLE RIVER BASIN

The Eagle River basin includes several gold bearing areas, each with their own unique history. The gold "rushes" into these areas were quite a bit later than the famous Pikes Peak Gold Rush of 1859.

Holy Cross City and the Holy Cross Mining District

This mining district didn't get started until a U.S. geologist named Ferdinand Hayden led a geological survey of western Colorado and parts of Utah. In his notes about the Eagle River, he noted that men fishing several creeks that ran south into the Eagle River had plenty of fish and that the men fishing the area were finding placer gold in the creek beds of Homestake Creek, Cross Creek, Beaver Creek, and Lake Creek. The initial publication of his maps in 1877 drew little attention.

However, in July 1880 the Leadville newspaper ran a story about a prospector who had come into town with a 60-pound piece of gold-rich ore which he said was from a quartz vein southeast of the Mount of the Holy Cross. Within a few weeks, other prospectors headed for the area. They quickly filed claims and organized a mining district which included the headwaters of both Homestake Creek and Cross Creek. Sadly, there wasn't enough gold in the placer deposits to earn decent wages, so the miners focused on the hard rock deposits. In the first year of real development, 1881, there was plenty of optimism. Holy Cross City was built at 11,335 feet elevation on the eastern slope of French Mountain and had a couple dozen buildings, with a reported 200 residents by 1882. The town of Gold Park was built four miles away at lower elevation and had about 400 residents. A flume was built between the towns with the idea that gold mined in the upper town would flow down to Gold Park for processing where there was more water.

By later in 1882 the mining was revealing a serious problem; the gold was only in shallow deposits. As the miners followed veins deeper, the gold simply ran out. By 1883, no new investments were being made and in all of 1884 the district produced only about 320 ounces of gold. Almost all of it came from the hard rock mines. In 1890, the whole excitement had faded away and the towns were both abandoned. One commentator wrote to the Leadville paper and said, of the placer deposits, "Several claims were worked, but none were ever found from which even wages could be taken out [let alone make a man rich]. The fishing... was far better than the placer mining."

If that doesn't discourage you, there is an opportunity to try your hand at catching a bit of the gold left behind by the original prospectors in this area; see the first couple sites in this chapter for details. Maybe bring a fly rod as well!

The history of <u>Fulford</u> (39.5151, -106.6558) includes interesting stories of prospectors dangling from ropes to collect cliffside mineral samples, town founder Arthur H. Fulford dying in an avalanche while on the way to post his mining claim on December 31st, 1891, and also many stories of "lost" mines in the area. The town was started in 1887 when prospecting started on East Brush Creek. It peaked at about 600 people but today has just two full-time residents; the rest of the private homes are used seasonally. The mining boom faded away in the first years of the 20th century as mines lost money. Today, the town itself is just a collection of private homes, not really worth a visit unless you are curious. A side note: I checked for placer gold at a public access spot near the town but didn't see any color at all.

On the other hand, the town of <u>Gilman</u> (39.5328, -106.3925) is pretty fascinating and sits in an amazing setting. It is on the flank of Battle Mountain, with great views of town from US-24 south of Minturn. The town was started in 1886 and in 1887 gold and silver were discovered in two vertical chimneys at the Ground Hog mine which then operated until the 1920s when it was killed by rising operating costs driven by inflation. The town was sustained by the discovery of the Eagle Mine which operated from the end of the 1880s through to the 1980s. The top layers of ore produced mostly gold but as the mine got deeper, the nature of the ore changed to sulfide ores which included much more silver, lead, copper and especially zinc. As they dug deeper, the silver and gold production reduced. Over the course of the years, the mine produced 393,000 troy ounces of gold along with all the other metals. Today the whole town is fenced off as a toxic EPA superfund site. However, there are great views of the town and mine from across the canyon on the highway. The mill is still in place as are many abandoned houses, shops and so on. There is gold in the Eagle River through here with public access points both upstream and downstream of the off-limits town.

Eagle River Prospecting Opportunities

The Eagle River gets its start on the North Side of Tennessee Pass. It flows north to Minturn to meet Gore Creek flowing out of the Vail area and then is accompanied by I-70 as it flows westward to meet the Colorado River just upstream of Glenwood Canyon. Most of the gold in this region is hard rock with very limited placers. However, visiting prospectors do have a chance to get their pan wet with some hope of finding color in quite a few spots. Gold enters the river near the headwaters with additional contributions from the surrounding countryside near the modern ghost town of Gilman and further downstream as mentioned in Hayden's survey, described above. There are prospecting sites and gold all the way west to the confluence of the Eagle and Colorado Rivers.

Site Number: T-01
Site Name: Gold Park Campground

This gorgeous USFS Campground on Homestake Creek in the Homestake Valley sits in a pretty, forested setting. PLO 1402 created a fairly large area here that is unclaimable and fun to explore.

Local Hints & Cautions:
- Bring your own water, there's no drinking water available at the campground. There is a public toilet.
- If there are anglers fishing, give them a wide berth.

Gold Finding Tips:
- Spend a minute thinking about how this creek changes during peak snowmelt. Where would the gold move then?
- The creek flattens out below the campground. Sample that area too. Gold drops out of a high flow where the surface flattens out.

Getting There: From I-70, Exit 171 for Minturn, head south on US-24 for 13 miles to FR-703, "Homestake Road". Turn right onto this good dirt road and drive 7 miles to the campground on the left at 39.4038, -106.4357.

Locale: rural Eagle County between Minturn and Tennessee Pass
Land Type: medium sized creek running through a narrow mountain valley
Land Manager: USFS – White River National Forest

Boundaries: From upstream by the campground 39.4031, -106.4359 to downstream at 39.4160, -106.4226. Since the campground is at one end of the recreational access area, you may even want to drive the road to find your favorite spot. For example, there is a great spot to pull off and park at 39.4134, -106.4302, which is where the creek diverges east, away from the road.

Key Regulations:
- Follow regulations posted at the campground.
- Remember not to park at a campsite unless you have reserved it.

Nearby Attractions & Accommodations:
The campground has 11 sites with fire grates and tables. Most sites can accommodate RV's up to 30-40 feet long.

Holy Cross Wilderness is nearby. The jeep road to Holy Cross City is barely passable to 4WD vehicles with winches. For most visitors, walking the four miles to see the historic ghost town would be the wiser choice. To find the trail, look just uphill of the campground for a dirt road labeled FR-759.

Homestake Reservoir is three miles further west on Homestake Road. Canoes, row boats, and small motorboats are allowed on the modest sized reservoir. Fishing is said to be "fair to good" in the reservoir and in the creek downstream.

Site Number: T-02
Site Name: Blodgett Campground

This rather informal dispersed USFS Campground is on a flat area next to Homestake Creek. The creek is across the road from the campground if that isn't obvious! PLO 1378 made this area unclaimable.

Local Hints & Cautions:
- There are no facilities at this campground.

Gold Finding Tips:
- Some parts of Homestake Creek here are more mud rather than cobbles, the gold is likely to accumulate among the cobbles. That's what I saw.

Getting There: From I-70, Exit 171 for Minturn, head south on US-24 for 13 miles to FR-703, "Homestake Road". Turn right onto this good dirt road and drive 1/4 mile to the campground on the left at 39.4725, -106.3664.

Locale: rural Eagle County between Minturn and Tennessee Pass
Land Type: variable sized creek running through a narrow mountain valley
Land Manager: USFS – White River National Forest

Boundaries: Upstream: 39.4719, -106.3711 to downstream at 39.4744, -106.3687, which includes about 1/3 mile of creek.

Key Regulations:
        Follow regulations posted at the campground.

Nearby Attractions & Accommodations:
This campground is dispersed camping, which means no services are provided. Site T-01 is further up the same road and offers pit toilets.

Site Number: T-03A, B
Site Name: Hornsilver Campground

This small USFS Campground is on a hillside next to the highway with Homestake Creek on the other side of the road. You'll pass this spot if you are headed to T-01 or T-02. PLO 1605 removed this prospecting area from claimable status.

Local Hints & Cautions:
- Bring your own water, there's no drinking water available at the campground. There is a public toilet.
- There is a pull-off with space to park a few vehicles on the highway just across the road from the campground. This would be a good day-use parking choice.
- This access includes almost a mile of creek so you may want to try both access points.
- Give any anglers a wave and a wide berth out of mutual respect.

Gold Finding Tips:
- This is a tricky area. During periods of high water, this area of the creek becomes a marsh with threads of creek running all over the property. Look for deposits of river cobble and gravel to find the gold.
- The actual mainstream of the creek is fairly far from the road. Late in the season, the main creek is the only water.

Getting There: From I-70, Exit 171 for Minturn, head south on US-24 about 12 miles to:
T-03A: the campground at 39.4892, -106.3676; visitors can also avoid the campground fees by parking on the west side of the road, just across from the campground.
T-03B: a long loop which lets visitors drive closer to the creek. The loop connects to US-24 at both 39.4945, -106.3679 and further downstream at the junction with Peterson Creek Rd. at 39.4960, -106.3689

Locale: rural Eagle County between Minturn and Tennessee Pass
Land Type: variable sized creek running through a narrow mountain valley
Land Manager: USFS – White River National Forest

Boundaries: Upstream: 39.4864, -106.3695 to downstream at 39.4955, -106.3699 at Peterson Creek Rd. Since the creek runs south to north, you can just keep an eye on the latitude to know you are legal.

Key Regulations:
- Follow regulations posted at the campground.

Nearby Attractions & Accommodations:
The campground has just 7 sites with 30' parking at each. The overall feel of this campground is cramped.

Site Number: T-04
Site Name: Minturn Boneyard Open Space

This is sort of a weird park but, hey, it has gold so whatever right? The whole section of river here was reacquired by the U.S. government so it is unclaimable.

Local Hints & Cautions:
• Avoid the spring melt.

Gold Finding Tips:
• Big boulders in the river provide easy spots to find color.
• Walking a little way upstream may get you to virgin ground.

Getting There: Hop off of I-70 at the US-24/Minturn exit. Drive south on US-24 through Minturn and turn east at 1352 US-24 to park at 39.5767, -106.4132 where you will find parking for about 10 vehicles.

Boundaries: Upstream access starts well upstream of the parking lot at 39.5745, -106.4073 (focus on the latitude for this boundary) and extends a little bit downstream of the parking area to 39.5775, -106.4144 (focus on the longitude for this boundary). There is 4/10 of a mile of river access here.

Locale: Minturn
Land Type: river running through a high mountain valley with some development nearby
Land Manager: Town of Minturn & Eagle County Open Space, USFS

Key Regulations:
• No gas-powered equipment in the open space park.
• Follow all posted rules.

Nearby Attractions & Accommodations:
Check out the town of Minturn. Definitely stop for a beverage at the Minturn Saloon. See my comments on the next site.

Site Number: T-05
Site Name: Minturn Little Beach Park

This cute park provides access to more river than it initially looks like. It is conveniently located right next to the town of Minturn.

Local Hints & Cautions:
- People's homes back up to the river so assume you are being watched. Play nice!

Gold Finding Tips:
- The gold here is fairly small but also fairly easy to find.

Getting There: From I-70 Exit 171 south on US-24 to the south end of town, turn east onto Cemetery Road, follow signs to the right to 39.5812, -106.4223 to park.

Boundaries: Upstream access starts at 39.5799, -106.4194 and continues to 39.5822, -106.4239. Do not leave the bed of the river on the far side as it is all private property or the shoulder of the federal highway.

Locale: Minturn, between Vail and Eagle
Land Type: very small river
Land Manager: Town of Minturn

Key Regulations:
- Only dig in the river.
- Pans and sluices only.
- Fill all holes and leave things as you found them.

Nearby Attractions & Accommodations:
The town of Minturn is cute, take a little time to explore. The Minturn Saloon is a favorite of mine. A bit pricey but exceptional food and atmosphere. The old bar is from the 1830's and was already about 50 years old when it was moved here from back east!

Site Number: T-06A-D
Site Name: Dowd Chute

This site really ends with Dowd Chute and the confluence of Gore Creek and the Eagle River, which is to say the prospecting area reaches quite a distance upstream as well. There is a ranger station here at the boat ramp, so if you have questions about other recreational activities or camping options, now is the time to ask!

Local Hints & Cautions:
• Do not dig around or otherwise block the boat ramp.
• Stay away from any anglers you see when you arrive on site. They need quiet (ok, due to the road noise, it isn't really quiet here but I mean no human voices) to avoid spooking the fish.

Gold Finding Tips:
• The area around and just downstream of the confluence should be interesting to prospect (although I haven't yet).
• The bulk of this site is a long inside bend. Sounds like a great spot for gold prospecting!
• The area just downstream of modern bridges tends to produce better gold due to the gold kicked up off of bedrock during the construction process... check that are out here.

Getting There: Hop off I-70 at Exit 171 and pick your spot to park from a variety of pull-offs within sight of the exit:
T-06A: Near the upstream end at 39.6002, -106.4359.

T-06B: 39.6025, -106.4388 on the north shoulder of the road.

T-06C: Parking by the boat ramp at 39.6073, -106.44455, or across the street at Meadow Mountain parking next to the Holy Cross Ranger Station.
Note: the confluence is between these two parking sites.

T-06D: Near the downstream end at the "Top of the Rockies" scenic byway sign at 39.6098, -106.4525 on the right side of the road when heading west. Or on the shoulder, just across the road, slightly to the east. Be careful getting down to the water here!

Boundaries: Upstream access starts a long way upstream of the boat ramp and confluence at 39.5975, -106.4339. Downstream access ends just below the confluence at 39.6104, -106.4531 where the river finally gets far enough away from US-6 to lose its unclaimable status. These boundaries mean we have just over 1.4 miles of prospecting to get on with.

Locale: Just north of Minturn along US-24 and US-6/I-70
Land Type: river running alongside noisy highways
Land Manager: USFS

Key Regulations:
- No power equipment within 50 feet of the highways. So, basically, no power equipment at all.
- Avoid any action which would damage erosion control or other structures.

Nearby Attractions & Accommodations:
Minturn just to the south on US-24, Vail to the east on I-70, Eagle-Vail just downstream to the west.

Site Number: T-07
Site Name: "The Bus Stop"

Yup, it's really a bus stop. Well, just west of the bus stop is a small parking area and a set of stairs down to the river. Getting down to the river along here can be a real challenge so the stairs are a true blessing.

Local Hints & Cautions:
- Be careful on the stairs, they aren't in perfect condition by any means.
- Stay out of the way of anglers who get there before you. The good news is they will move along quickly in most cases!

Gold Finding Tips:
- Hike along the riverbank upstream to the inside bend where I easily found color in my pan.
- Like many sites along here, it is easy to find a little fine gold. Coming when water levels are low will make the search easier and access to the full length of this area safer.

Getting There: The bus stop is just off of US-6. Turn off of the main road onto Kayak Ct. at 39.6181, -106.4629 and take a quick left to the bus stop. The parking is just west of the actual bus stop. The stairs are right in front of the parking area.

Boundaries: Upstream access starts just downstream of the bridge at 39.6179, 106.4604, 800 feet downstream to the stairs and then another 760 feet downstream to 39.6194, -106.4658.

Locale: Eagle River running just north of US-6 & I-70
Land Type: river running through a mountain area
Land Manager: State of Colorado/CDOT

Key Regulations:
- See prior site for details.

Nearby Attractions & Accommodations:
Beautiful scenery, Vail, Minturn, Eagle, other dig sites...you get the idea.

Site Number: T-08
Site Name: West Edwards River Access

A simple pull-off from US-6 here provides parking for several cars and decent access to the river. The long inside bend here means there is gold to be had!

Local Hints & Cautions:
- This site is just downstream of a highway bridge. Be sure to sample near the bridge since bridge construction often brings up gold from bedrock. All that material is typically just dumped in the river as the construction progresses.

Gold Finding Tips:
- It's fairly easy to find color in your pan here...and fishing weights. (Please remove any fishing weights and dispose of them properly.)

Getting There: Parking is just off of US-6 at 39.6757, -106.6467.

Boundaries: Upstream access starts just upstream of the highway bridges at 39.6731, -106.6470 and extends downstream to the end of the inside bend at 39.6779, -106.6470; both boundaries are at the edges of private property.

Locale: Eagle River just west of Edwards
Land Type: river running through a rural area
Land Manager: BLM

Key Regulations:
- No power equipment within 50' of the highway bridge.

Nearby Attractions & Accommodations:
The town of Edwards is just to the east. If you are a golfer, Eagle Springs Golf Club is just to the west on US-6.

Site Number: T-09A-C
Site Name: Wolcott BLM CG

This campground offers picnic tables, charcoal grills, toilets and great river access on an inside bend, a boat ramp, but also a day-use fee. If you aren't camping here (there are only 6 spots) and want to dodge the day use fee, there is also a pull-off just east of the campground as detailed below. This is a very popular fishing area.

Local Hints & Cautions:
- No gas-powered equipment at the campground. No restrictions outside of that area.
- Avoid the boat ramp when prospecting.

- Avoid disturbing anglers by giving them some space.

Gold Finding Tips:
- Use the informal trails created by the anglers to access multiple spots along the river.

Getting There: US-6 west of CO-131 & Exit 157, from exit go north to frontage road, west 2 miles past CO-131 bridge and "Wolcott Yacht Club" to all three access points.

T-09A: 39.7108, -106.6940 turn right to dirt parking. Do not block the way for other vehicles (Eagle River Water & Sanitation maintains a well here). This dirt lane runs almost all the way to the upstream edge of the prospecting area.

T-09B: at 39.7118, -106.6959 turn right into campground area.

T-09C: Another informal pull off near the downstream end of the area at 39.7123, -106.7022.

Boundaries: Upstream access starts all the way upstream at the start of the inside bend at 39.7058, -106.6887. Downstream the limit is past the campground at 39.7124, -106.7024.

Locale: rural Eagle County just west of Wolcott
Land Type: river running through a dry rural valley
Land Manager: BLM, Denver County (weird right?!), Eagle River Water & Sanitation

Key Regulations:
- Respect any signage at T-09A; this property is owned by the City & County of Denver and by Eagle River Water & Sanitation. No gas-powered equipment at the campground.
- Follow posted rules in the campground. Don't park in a campsite unless you have paid for it. No gas-powered gear in the campground.

Nearby Attractions & Accommodations:
Camp onsite here. Also rumors of great trout fishing have been heard.

Site Number: T-10
Site Name: Mott's Landing

Another easy access point with lots of parking, just a bit west of the last, but no user fees here. Also provides access to the river just downstream of a major I-70 bridge.

Local Hints & Cautions:
• Digging in the shade of the bridge sounds pretty good on a hot summer day.

Gold Finding Tips:
• Check out the "usual spots" downstream of the highway bridges.

Getting There: Use the driving instructions from the prior site and continue west on US-6 to 39.7103, -106.7099; just west of the I-70 bridge this spot provides a nice wide parking area alongside US-6.

Boundaries: Upstream access starts just upstream of the bridges at 39.7113, -106.7087 and continues downstream along the inside bend to 39.7094, -106.7117.

Locale: Eagle River between Wolcott and Eagle
Land Type: river running next to the highway in a rural setting
Land Manager: CDOT, Eagle County

Key Regulations:
• See prior site for details.

Nearby Attractions & Accommodations:
Camping back at Wolcott BLM CG or head west to Eagle for lots of restaurants and other housing options.

Site Number: T-11A, B
Site Name: unnamed Eagle River Access

While I have highlighted a couple specific pull-offs in this stretch of river, feel free to use others if the access or the shape of the river seems good to you.

Local Hints & Cautions:
* There are some steep drops to the river here, be choosey about how you climb down and be careful!
* Avoid blocking the loop so others can use it as a loop.

Gold Finding Tips:
* The river here has lots of winding inside bends – fun to explore with your gold pan.

Getting There:
T-11A: pull off loop from US-6 is at 39.6950, -106.7413.

T-11B: 39.6887, -106.7485; a hard-to-spot pull-off just west of these coordinates has the best access.

Boundaries: Upstream access starts at 39.7000, -106.7310, and continues to 39.6860, -106.7517 for a total of about 1.5 miles of river.

Locale: Eagle River between Wolcott and Eagle
Land Type: medium sized river in a mountain valley
Land Manager: BLM

Key Regulations:
* No powered mining equipment within 50' of the road.

Nearby Attractions & Accommodations:
See prior site.

Site Number: T-12
Site Name: Eagle County Fairgrounds

The fairgrounds, baseball diamonds and county roads storage area add up to a lot of land along the Eagle River.

Local Hints & Cautions:
• None noted.

Gold Finding Tips:
What's not to love the big bend in the river here?

Getting There: From I-70 Exit 147, head south through the roundabouts there to the roundabout with Chambers Ave. Take the first exit, for Chambers Ave. to head westerly. After a few hundred feet, the road curves right and then left to become Fairgrounds Rd. The first left leads to the Visitor Center with the actual fairgrounds a little further west. Drive to either the parking and river access at 39.6525, -106.8368 which is near the upstream edge of the interesting area or a bit further to the paved lot at 39.6516, -106.8402 which is more toward the middle.

Boundaries: The upstream boundary is the downstream edge of the kayaking area. The downstream boundary is a long way down around the inside bend to 39.6484, -106.8517 beyond which there is private property.

Locale: in the town of Eagle
Land Type: medium-sized river flowing through town, near the interstate
Land Manager: Eagle County, Town of Eagle

Key Regulations:
• Respect all posted rules.
• No powered equipment of any sort.
• Stay away from the boat ramp except to use it as access down to the river.

Nearby Attractions & Accommodations:
Camping back at Wolcott BLM CG or here in Eagle for lots of
restaurants and other housing options.

Site Number: T-13
Site Name: Brush Creek Open Space Park

Brush Creek is one of the little waterways that contributes a
bit of gold to the Eagle River. This park has a wide variety of
facilities and uses. This site is on Brush Creek, not the
Colorado River.

Local Hints & Cautions:
- Stay out of the manicured areas of the park.
- Play nice, clean up after yourself, be a good ambassador for
  gold prospecting.

Gold Finding Tips:
- Little creeks like this are quirky. Sample, sample!

Getting There: From I-70 Exit 147, head south through the
roundabouts following Eby Creek Rd. to the roundabout with
US-6/Grand Ave. Take the first exit, for Grand Ave. to head
westerly. After a few hundred feet, turn left onto Capitol St.
and follow that to the Brush Creek Park and Pavilion parking
lot at 909 Capitol St.

Boundaries: From Sylvan Lake Rd at 39.6378, -106.8181, both
upstream and downstream of Capitol St. and everything down
to Sylvan Lake Rd. again at 39.6464, -106.8376 (as the road
curves around toward US-6). The parking is off of Capitol St. at
39.6439, -106.8297.

Locale: in the town of Eagle
Land Type: mountain creek flowing through a meadow
Land Manager: Town of Eagle

Key Regulations:
- Respect all posted rules.

- No powered equipment of any sort.

Nearby Attractions & Accommodations:
Camping back at Wolcott BLM CG or here in Eagle for lots of restaurants and other lodging options.

Site Number: T-14
Site Name: Brush Creek Confluence Open Space

Brush Creek is one of the little waterways that contributes a bit of gold to the Eagle River, and this is where the two flows meet. This park gives access to the Eagle River via a trail from the parking area along the river.

Local Hints & Cautions:
- Only use established trails.

Gold Finding Tips:
- The area just downstream of a confluence is often where the gold from both waterways drops.
- The inside bend before the confluence creates some interesting water flow and eddies, definitely worth sampling.

Getting There: From I-70 Exit 147, head south to the roundabout just south of the interchange. Take the first exit to Eby Creek St. from the roundabout and the second exit from the next roundabout to stay on Eby Creek St. After a quarter mile, exit the next roundabout at the first exit onto US-6/Grand Ave. One more time...exit the next roundabout at the first exit onto Violet Lane and follow that a tenth mile to 38 Violet Lane (39.6479, -106.8392) where a right turn will land you in the parking lot for the open space park. Follow the existing path to the river.

Boundaries: From 39.6484, -106.8408 past the creek to 39.6486, -106.8463. Once past the creek, stay within the banks of the river. The property is owned by the county but is an operating facility.

Locale: in the town of Eagle
Land Type: mountain creek flowing through a meadow
Land Manager: Town of Eagle, Eagle County

Key Regulations:
- Respect all posted rules.
- No powered equipment of any sort in the open space park.

Nearby Attractions & Accommodations:
Camping back at Wolcott BLM CG or here in Eagle for lots of restaurants and other housing options. The downtown area is cute, go check it out!

Site Number: T-15
Site Name: Trail Gulch

Trail Gulch road runs easterly from the Gypsum I-70 interchange and provides access to a stretch of river owned by the town. This area is fairly popular with local prospectors so visitors may get to chat with a local expert!

Local Hints & Cautions:
- Be sure to park where it is allowed (honor the rocks which mark the edge of the road on the river side).

Gold Finding Tips:
- There are lots of bends in the river with interesting cobble bars to explore here.

Getting There: Access via I-70 exit 140, US-6, south a block to the roundabout, the third exit is Trail Gulch Rd. Follow that to multiple parking spots along Trail Gulch or the larger lot at the end of the road.

Boundaries: Upstream access starts at 39.6510, -106.9350 which is out around the fishing lake. Downstream the limit is 39.6518, -106.9444.

Locale: on the east edge of the town of Gypsum
Land Type: highly braided river with lots of oxbows
Land Manager: City of Gypsum, CO Parks & Wildlife

Key Regulations:
- Parts of this area are owned by the state of Colorado and managed by CO Parks & Wildlife as Gypsum Ponds State Wildlife Area. They charge a user fee for day use. The way I have defined this site, we are only digging in the river on city property but crossing CFW land to get to the river is almost impossible avoid. If each visiting individual doesn't have a fishing or hunting license, an annual or day pass can be purchased at:
https://www.cpwshop.com/purchaseprivilege.page
by choosing the "Lands & Trails" option on the left menu.
- Camping, fires, glass containers, boat launching and takeouts prohibited. Also, dogs prohibited from March 15 to June 15 to protect nesting birds.

Nearby Attractions & Accommodations:
The town of Gypsum has lots to offer.

Site Number: T-16A-E
Site Name: Gypsum BLM CG

This campground is just west of the town of Gypsum. PLO 3843 made this campground and the large area downstream from it unclaimable.

Local Hints & Cautions:
- When the water is high, many parts of the river here flood over the inside bends. Be careful of ankle-breaking potholes in the ground caused by this repeated flooding.

Gold Finding Tips:
- The land here is rather flat and so the river has many bends in it. Lots to sample.
- Think about how the water would flow at peak volumes.

Getting There: I-70 Exit 140, go south to frontage road, west on frontage road for 1.5 miles past the roundabout to the left turn into the campground.

T-16A: The campground. It is on the south side of US-6 at 39.6556, -106.9761; there is excellent river access at both the far southeast side of the campground loop and on the west side, through the trees to the river from the day use area.

T-16B: west a bit from the campground on the frontage road to 39.6550, -106.9794.

T-16C: a bit further west and a larger parking area 39.6544, -106.9812.

T-16D: "Community River Access", further again and another large lot with good paths heading to two spots on the river 39.6503, -106.9845.

T-16E: "Horse Pasture River Access", one last spot with access to some crazy bends in the river 39.6498, -106.9923.

Boundaries:
The unclaimable area starts at the upstream end of the campground (39.6539, -106.9722) and extends to 39.6484, -106.9956 so it isn't surprising that along this 1.4-mile stretch of Frontage Road, there are multiple access points.

Locale: Eagle River west of Gypsum
Land Type: river running through in a high, dry mountain valley
Land Manager: BLM

Key Regulations:
- Follow rules on signage while at the campground.
- All types of equipment welcome outside of the campground as long as other users are treated with respect.

Nearby Attractions & Accommodations:
Fishing, river floating, camping onsite (first come, first serve),

bird watching (look for the bald eagles and blue heron!) Visit the volcano crater and lava flow just to the west of here near Dotsero.

Site Number: T-17
Site Name: Aunt Sara's Riverdance RV Resort

This large RV park owns a large stretch of river and allows guests to prospect while staying there.

Local Hints & Cautions:
• Follow any advice from your hosts.

Gold Finding Tips:
• If you want to dig right behind your campsite, pay for a riverside site. I tested a couple of them, and they do have gold.
• If you want good gold, you probably want to explore the big bends in the river.
• Take a walk with your pan on the hiking trail that runs across the property from the west edge of the formal resort area.

Getting There: First, make a reservation by calling 970-400-7078 or going to riverdancepark.com; then take I-70 to Exit 140. Head south from the interchange to the roundabout, taking the first exit to US-6/I-70 Frontage Rd. and follow that west for 2.7 miles. The address is 6700 US-6, Gypsum, CO 81637 The GPS is 39.6482, -107.0029.

Boundaries: Upstream access starts well before the formal RV park at 39.6485, 106.9957 and extends well past the formal park to 39.6473, -107.0145; yup, that's right, these folks own a mile of the river!

Locale: in the canyon just west of Gypsum
Land Type: river with multiple oxbows
Land Manager: the RV Resort, this is private land

Key Regulations:
- Follow the rules shared with you at check-in.

Nearby Attractions & Accommodations:
If you are digging here, you are staying here.

Site Number: T-18
Site Name: Eagle County River Access

Just a stone's throw west of the prior site, this undeveloped county site doesn't offer much, well, except a big fat inside bend. We don't really need anything more, do we?!?

Local Hints & Cautions:
- Drive in a little way and turn left into the main parking lot or further toward the river.

Gold Finding Tips:
- The inside bend here has a classic shape that just screams gold accumulation to me at the head of the bend. I hope you find out that I am right!

Getting There: Follow the instructions to the prior site and then continue west on the frontage road to 39.6464, -107.0171 and turn left. Be sure to avoid the private driveway just east of this turn.

Boundaries: Upstream access starts at 39.6460, -107.0159 and continues around the big bend to 39.6444, -107.0229.

Locale: western Eagle County, west of Gypsum
Land Type: river running through an arid canyon
Land Manager: Eagle County

Key Regulations:
- None noted. Follow any posted rules.

Nearby Attractions & Accommodations:
Camp at one of the prior sites. FYI, the nearest public dump

site for RVs is in Edwards.

Site Number: T-19
Site Name: BLM River Access

Another inside bend? This river just keeps on giving! A big thanks to the rafters who established this informal boat ramp originally.

Local Hints & Cautions:
• Don't mess with the boat ramp.

Gold Finding Tips:
• It's a big inside bend. Let the fun begin!

Getting There: I-70 to Exit 133 then, along the south side of the freeway, US-6 to 39.645, -107.034 and turn south onto the property. Drive or walk several hundred yards from the pull-off to the river. Coming from the prior site, just follow US-6/frontage road west to the GPS coordinates.

Boundaries: Public access from upstream at 39.643, -107.033 to downstream at 39.642, -107.036 with a big inside bend between those points.

Locale: western Eagle County, west of Gypsum toward Dotsero
Land Type: river running through an arid canyon
Land Manager: BLM

Key Regulations:
• None noted.

Nearby Attractions & Accommodations:
Look for accommodation at the prior sites if camping, in Dotsero or Gypsum if a hotel is your style.

From here, the Eagle river flows into the Colorado River. The next sites downstream are on the Colorado River. They are in Chapter S, in the section "The Colorado River along I-70/US-6".

Kevin Singel

Sampling the Poudre River

Poudre River Sampling: I think I saw a speck in this pan somewhere

CHAPTER U: POUDRE RIVER

Prospectors in Fort Collins, Loveland and the northern front range who are looking for local prospecting are likely to be somewhat disappointed. However, their pans won't be entirely empty. Lots of people asked for this info, so here it is for you.

Prospecting Opportunities

The Poudre River is designated "Wild and Scenic" in the canyon upstream of town so only panning is allowed there. The main gold bearing areas are actually just north and south of the upper Poudre River in the mountains. Those creeks lack public access due to private property and mining claims. However, once these tributaries feed into the Poudre, our odds of finding gold improve a bit. Since these creeks meet the Poudre just about as it reaches the flat lands, our best access, and the best gold (still not very good, to be candid) are down on the plains in Larimer County. Thanks to Larimer County and its sensible commissioners, we get to do a little prospecting in their open space parks as long as we are respectful.

NOTE: Some rangers are fine with prospecting in the riverbed, others have said the 2024 rules forbidding removal of anything from the Open Space also apply to the riverbed. Prospect at U01 thru U-03 at your own risk understanding some rangers may chase you off.

Site Number: U-01
Site Name: Lions Open Space Park

Lions Open Space Park is part of the Larimer County Open Space system of parks and is braided through by the Cache La Poudre River. The river is not particularly known for prospecting but there is some very fine gold to be found. The river also drains an area known for kimberlite pipes and there has been commercial diamond mining in the Poudre River drainage so finding diamonds is also a possibility (in theory at least).

Local Hints & Cautions:
- Late April through early June is the peak runoff period when prospecting here can be difficult or even dangerous since the Poudre River is undammed.

Gold Finding Tips:
- The gold here is extremely fine. In my sampling, everything was -150 to 200 mesh, which is about as small as my eyes can see individual specks of gold in bright daylight.
- Bring your heavy panning concentrates home for final careful processing. This gold is so small that you will find a little more hiding in the black sands if you process it carefully at home.
- Check out the inside bend on the west side in the middle of this property and the one on the easy, east side near the parking area.

Getting There: From I-25, Exit 269B & follow directions to 2425 N Overland Trail, Laporte. You will see a park sign at the entrance to the gravel lot and toilet facilities just past the fence on your right as you enter the parking area. From the parking lot, walk over to the recreational path and pick your creek access trail.

Locale: Laporte in Larimer County.
Land Type: medium sized river running through flood plain
Land Manager: Larimer County Open Space

Boundaries: Upstream from 430.6245, -105.1428 with access to both sides of the river to 40.6221, -105.1404 where access becomes limited to the east bank, to downstream at the N Overland Trail road bridge where access ends.

Key Regulations:

- Respect signage and fences.
- Do not dig near the bridge or other improvements.
- No gas-powered equipment.
- NO digging into the slope from the rec path down to creek level.

Nearby Attractions & Accommodations:
Little Laporte is nearby and just south of here is Fort Collins' Old Town which is loaded with bars, restaurants, and shops. There are private campgrounds such as a KOA fairly nearby or head west into the hills to camp or stay in a lodge.

Site Number: U-02
Site Name: Shields River Access

This park provides river access for a variety of users. There is a rec path on the far side of the river which can accessed via a sidewalk along the west edge of the Shields St. bridge at the south end of this property.

Local Hints & Cautions:
- See points included in the prior site.

Gold Finding Tips:
- The Cache Le Poudre River Trail offers plenty of access to the river through this park.

Getting There: 1303 N Shields St., with public parking west off of Shields St at 40.6044, -105.0957.

Boundaries: Upstream at 40.6054, -105.100 down to the N. Shields St. bridge.

Locale: northern Fort Collins
Land Type: river running through a semi-rural area
Land Manager: Larimer County

Key Regulations:
- See prior site for details.

Nearby Attractions & Accommodations:
There's lots of fun stuff and places to stay in Fort Collins. Of course, Fort Collins is a craft-brewing destination and Colorado State University offers fun cultural and sports events.

Site Number: U-03
Site Name: River Bluffs Open Space Park

This large park and open space area has river access, fishing ponds, toilets, and drinking fountains. It is southeast of Fort Collins and just west of Windsor.

Local Hints & Cautions:
• See points included in the prior site.

Gold Finding Tips:
• The Cache la Poudre River Trail runs along here providing access to lots of river exploration.

Getting There: Head to 6371 E CR-32 E, Windsor to find this park (40.4865, -104.9583) where there is a large parking lot. The river is right at the south edge of the parking lot.

Boundaries: Upstream access starts at CR-32E (40.4865, -104.9622) and ends at CR-32 (40.4794, -104.9535) when the trail continues onto HOA property.

Locale: southeast of Fort Collins, just east of I-25 Exit 262
Land Type: river running through a rural area
Land Manager: Larimer County

Key Regulations:
• See prior site for details.

Nearby Attractions & Accommodations:
The town of Windsor is just to the east.

Site Number: U-04
Site Name: Poudre Learning Center

Greeley doesn't allow gold prospecting in their open space parks but they do allow prospectors to cross the park lands to access the Cache la Poudre River in certain areas. If you are crazy enough to prospect here, you have my admiration! This is not a place to expect much.

Local Hints & Cautions:
*   Park so you are not causing trouble for the residents of the neighborhood.

Gold Finding Tips:
*   The Poudre Trail runs along here providing access to lots of river exploration.

Getting There: The Learning Center provides parking at 8313 W F St., Greeley but it's a longer walk to the river from there. I suggest parking in the neighborhood at one of the following points that provide access to the Poudre River Trail: River Run Drive at 40.4414, -104.8071 or just east of there on Poudre River Rd. at 40.4409, -104.8044.

Boundaries: Upstream access starts at 40.4463, -104.8165 and extends along the river to 40.4441, -104.7973.

Locale: Greeley
Land Type: river running along the edge of a city
Land Manager: Greeley Parks

Key Regulations:
*   No digging in the open space parks, only in the river off of "park" lands.

Nearby Attractions & Accommodations:
The whole town of Greeley. Sites in Chapter A for the South Platte River near here such as the site in Evans.

CHAPTER V: TRINIDAD & THE PURGATOIRE RIVER

Prospectors all the way south on the Colorado Front Range, in Trinidad, may be surprised they have some interesting opportunities to prospect locally.

The Purgatoire River has its origin in the Spanish Peaks of the San Isabel National Forest and flows downhill to Trinidad and then northeasterly across the plains to its confluence with the Arkansas River on the edge of the town of Las Animas, just west of John Martin Reservoir. The name was given to the river by French explorers; in English it would be Purgatory River as you might guess. Hopefully, those who visit agree it's not purgatory but rather a little slice of heaven!

The little city of Trinidad sits along the Purgatoire River just downstream of Trinidad Lake State Park. Despite having only around 8-9,000 residents, it is the largest town in Las Animas County. It is just eleven miles north of the Colorado/New Mexico border.

Side Trip: One quirky-cool site to see here is to see the geological layer called the K-T Boundary. This thin, dark layer is visible in scattered spots all over the world. It was created by the dust cloud which formed when a meteor hit the earth near the Yucatán Peninsula of Mexico. This was the event that killed the dinosaurs. Here's how to see it: Visit Long Canyon Trail at **Trinidad Lake State Park**, County Rd. 18.3, Cokedale, CO 81082; it is at 37.1215, -104.6049; the actual hike is an easy ¼ mile. Ask for details at the state park visitor center.

Prospecting Opportunities

The Middle Fork of the Purgatoire River is gold bearing. There are modern reports of gold as far upstream as Stonewall, but that area is locked up in private property. If you have a friend with property on the river, go check it out. Otherwise, try the following sites a bit downstream.

Site Number: V-01
Site Name: County Land above the reservoir

An odd patch of undeveloped land, I imagine this area being turned into a more developed open space park in the future. This spot, a bit west of Trinidad Lake State Park offers us access to the Purgatoire River upstream of the reservoir.

Local Hints & Cautions:
- I have not visited this site so just keep your eyes open for signage or other uses of this land which you should respect.

Gold Finding Tips:
- Expect the gold to be very small particles.

Getting There: From I-25 Exit 14B onto W. Colorado Ave. and follow it to San Juan St. Turn left and follow the curve right onto Robinson Ave./CR-12. From this point the access area is about 8 miles westerly on CR-12. From CR-12 at 37.1309, -104.6336, pull off the road on the south side and follow the two-track dirt route east to 37.13106, -104.6308 where the county land starts.

Boundaries: Access the river due south of 37.13106, -104.6308 and to the east until the downstream boundary at longitude -104.6267.

Locale: Las Animas County, west of Trinidad
Land Type: river running through an arid valley
Land Manager: Las Animas County

Key Regulations:
• None noted; follow any posted signage.

Nearby Attractions & Accommodations:
Check out the cute small-town vibe of downtown Trinidad. Enjoy the reservoir.

Site Number: V-02A, B
Site Name: Trinidad Lake State Park

The river just upstream of the actual reservoir is on state park property for a decent distance. This site highlights a couple different ways to conveniently access the river. Since this area is part of a state park, it is important to respect the strict rules of the state park system. Please refer to the rules section below.

Local Hints & Cautions:
• I have not visited this site so just keep your eyes open for signage or other uses of this land which you should respect.

Gold Finding Tips:
• Expect the gold to be very small particles.

Getting There: From I-25 Exit 14B onto W. Colorado Ave. and follow it to San Juan St. Turn left and follow the curve right onto Robinson Ave/CR-12. Then:

V-02A: Pull off of CR-12 on the south side of the road at 37.1309, -104.6336 and follow the dirt double track to the end near 37.131, -104.626. From there walk south to the river. The state park abuts the county land so as soon as you are east of the county land, longitude mentioned in the prior site, you are on state park lands.

V-02B: Turn left onto CR-57.7 at 37.1427, -104.6172 and drive about 3.2 miles south and east into Riley Canyon to pull-offs like 37.1328, -104.6050 or other spots along the way to the east, all the way out to the east end of the road, depending on how full the reservoir is at the time of visit.

Boundaries: From longitude -104.6267, east to the pool of the reservoir.

Locale: Las Animas County, west of Trinidad
Land Type: river running through an arid valley
Land Manager: Colorado State Parks

Key Regulations:
- Regulations in state parks are quite strict and must be respected. Individual rangers also have the right to tell you leave...or worse if they see something they do not approve of.
- Pans and shovels only; no removal of concentrates is allowed. Gold prospecting in a state park is encouraged as a recreational activity only; leave your dig area exactly as you found it by filling any holes, smoothing out tailings, etc. Also, if you find anything interesting, like a nugget, they ask you to show the rangers. It belongs to the state but I'd request a receipt and ask what they intend to do with it. If it was me, I'd be happy to hand it over after I take pictures because I'd hope they display it in their visitor center!

Nearby Attractions & Accommodations:
Check out the cute small-town vibe of downtown Trinidad.

Site Number: V-03
Site Name: Boulevard Loop Trail

Even though Trinidad is a fairly small town, the town owns some nice area along the river. This undeveloped site is still a bit of a gem with a half-mile loop trail, picnic tables, lots of shade from old cottonwood trees, and for us, a long inside bend to prospect.

Local Hints & Cautions:

- I have not visited this site so just keep your eyes open for signage or other uses of this land which you should respect.

Gold Finding Tips:
- Expect the gold to be very small particles.

Getting There: From I-25 Exit 14B onto W. Colorado Ave. and follow it to San Juan St. Turn left and follow the curve right onto Robinson Ave./CR-12. Turn onto Alta St off of Highway 12. Follow Alta St as it curves left to become Boulevard St. Follow that to the end of the road and the trailhead 37.16475, -104.5160.

Boundaries: Access the river from the Railroad bridge over the little creek at the upstream end of this giant inside bend, downstream to the I-25 bridges which mark the beginning of the next site. To access the northwest side of the river past the interstate bridges, continue walking along the bank of the river to any spots that look interesting as far downstream as the College Ave. bridge.

Locale: Trinidad
Land Type: river running along the edge of a developed city
Land Manager: City of Trinidad

Key Regulations:
- Follow any posted signage.
- No gas-powered equipment due to all of the road and bridge work along here and the city open space rules.

Nearby Attractions & Accommodations:
Check out the cute small-town vibe of downtown Trinidad. Enjoy the reservoir.

Site Number: V-04A-C
Site Name: Trinidad Riverwalk

A recreational path along the river provides access to about 3.5 miles of waterway along the south bank of the Purgatoire River through Trinidad. The path is actively used for fishing and tubing.

Local Hints & Cautions:
• Respect the other users including the tubers and anglers.

Gold Finding Tips:
• Expect the gold to be very small particles.

Getting There: There are several easy access points to the trail.

V-04A: 1 Next to I-25 and a nice inside bend at 639 W Main where there is public parking behind the Exxon gas station.

V-04B: City Park, 427 N Chestnut St., has a playground.

V-04C: Waggin' Tails Dog Park, at the corner of N Linden Ave. and E Elm St. (37.172, -104.498).

Boundaries: The river adjacent to the rec path from its start across from the Days Inn & Suites (on the rec path at 37.161, -104.515) downstream to the Dog Park site with the actual "on river" boundary at 37.173, -104.500 which is just a bit downstream of the Linden St. bridge.

Locale: in the city of Trinidad
Land Type: river running through town in an arid plain
Land Manager: City of Trinidad

Key Regulations:
• Follow any posted signage.
• No gas-powered equipment due to all of the road and bridge work along here and the city open space rules.

Nearby Attractions & Accommodations:
Check out the cute small-town vibe of downtown Trinidad. Enjoy the reservoir.

CHAPTER W: ROARING FORK RIVER BASIN

While the Aspen area was built on silver and isn't known for gold at all, the Roaring Fork River does have some gold in it. The Independence Pass District, on the upper Roaring Fork River produced the bulk of the 28,200 ounces commercially produced from Pitkin County. The town of Independence is mentioned in the prior guidebook *Finding Gold in Colorado: Prospector's Edition,* so it's not included here but suffice to say Independence is the first, but not last, area that contributes gold to the Roaring Fork River.

A special thanks to members of the Finding Gold in Colorado Facebook group for giving me some tips on where they found gold in this watershed. Their help meant I could get this book out faster, without having to test these sites myself. Also, a shout-out to all the towns along the river who welcome casual gold panning in their town.

> **Site Number: W-01**
> **Site Name: Lost Man CG**

Just off of CO-82 and a bit downhill from the high alpine ghost town of Independence, this USFS Campground is the first chance to prospect in the Roaring Fork River as we head downstream. This campground is in a heavily wooded area.

Local Hints & Cautions:
- This area is heavily wooded; don't lose track of your directions, it would be easy to get turned around and lost.

Gold Finding Tips:
- I haven't prospected this area yet.

Getting There: The campground is just off of CO-82 to the south at 39.1217, -106.6245; those not staying in the campground can park across the street or pull off on the shoulder closer to the center of the prospecting area at 39.1225, -106.6270 if that looks like a good fit for your vehicle.

Boundaries: Upstream access starts upstream of the campground at 39.1193, -106.6235. Downstream access is quite a bit west of the campground at 39.1226, -106.6294 for just over ½ miles of river access.

Locale: high in the forest southeast of Aspen
Land Type: river running through a modest valley
Land Manager: USFS

Key Regulations:
- No gas-powered equipment near the campground.

Nearby Attractions & Accommodations:
Stay here or at one of the following sites at other campgrounds.

Site Number: W-02
Site Name: Lincoln Gulch CG

This dispersed campground (meaning no services) is just south of the river.

Local Hints & Cautions:
- This area is heavily wooded; don't lose track of your directions, it would be easy to get turned around and lost.

Gold Finding Tips:

- I haven't prospected this area yet.

Getting There: The campground is off CO-82, about 20 minutes east from Aspen. From CO-82, turn south onto Lincoln Creek Rd. (FR-23) and turn right toward the campground at the fork in the road 0.4 miles in. Another ¼ mile west is the campground. The last ½ mile is on a poorly maintained forest service road.

Boundaries: Upstream access starts at the east edge of the campground at 39.1180, -106.6953. Downstream access is quite a bit west of the campground at 39.1188, -106.6994 for just over ½ miles of river access.

Locale: high in the forest southeast of Aspen
Land Type: river running through a modest valley
Land Manager: USFS

Key Regulations:
- No gas-powered equipment near the campground.

Nearby Attractions & Accommodations:
Stay here or at one of the following sites at other campgrounds.

Site Number: W-03
Site Name: Difficult CG

Named for nearby Difficult Creek, this campground is definitely not difficult but some of the prospecting is!

Local Hints & Cautions:
- Much of the downstream portion of the river here is rather boggy and will be difficult to prospect during wet periods.

Gold Finding Tips:
- I haven't prospected this area yet.

Getting There: The campground is off CO-82, just a bit east of Aspen. From CO-82, turn south onto Difficult Campground Rd.

at 39.1489, -106.7817 and stay left toward the campground.

Boundaries: Upstream access starts at due south of the campground at 39.1383, -106.7725. The river all along the Difficult Campground road is included in the access area. Choosing a pull-off along the road may reduce walking distance and will definitely avoid a day-use fee. Downstream access is quite a bit west of the campground at 39.1459, -106.7811 for just over ½ miles of river access.

Locale: high in the forest southeast of Aspen
Land Type: river running through a modest valley
Land Manager: USFS

Key Regulations:
• No gas-powered equipment near the campground.

Nearby Attractions & Accommodations:
Stay here or at one of the other campgrounds along CO-82.

> **Site Number: W-04A-F**
> **Site Name: Aspen City Parks**

It may surprise some prospectors that fancy Aspen allows gold panning in their parks, but it's true! The rules are fairly strict but this a chance to have a little fun with a gold pan if you are in the area.

Local Hints & Cautions:
• Seeing a prospector is rare here and so you will be a novelty to the tourists and locals alike. Please be a friendly ambassador for our shared passion with anyone who is curious.

Gold Finding Tips:
• Go when water levels are low to help in spotting the gravel bars and larger rocks that will tend to accumulate the fine gold traveling downstream.

Getting There:
W-04A: Ute Park: Parking is at 39.1814, -106.8120 on Ute Ave. with off-street parking for about six cars. Follow the path from the parking area north to the river.

W-04B: Anderson Park: 1101 E Cooper Ave.

W-04C: Prockter and Herron Parks: These parks are across the street from each other on Neale Ave. with parking on both side of the street at 108 Neale Ave.

W-04D: Newbury Park, John Denver Sanctuary, and Rio Grande Park: This continuous stretch has designated street parking at 415 Rio Grande Pl., (39.1919, -106.8180).

W-04E: City facility: 590 N. Mill St. with plenty of parking.

W-04F: City park at the Aspen Institute: 1000 N Third St. with a large parking lot.

Boundaries: For all sites, stay on the park side of the river, the other side is generally private property.

W-04A: Ute Park: Upstream, straight north from the parking at 39.1822, -106.8118 to 39.1840, -106.8116 with a nice inside bend about halfway along.

W-04B: Anderson Park: From 39.1855, -106.8119 to the Cooper Ave. bridge.

W-04C: Prockter and Herron Parks: from 39.1892, -106.8130 and continuing into the next site.

W-04D: From the prior site downstream to 39.1931, -106.8174 with several nice bends and islands to explore.

W-04E: With an awesome inside bend from upstream at 39.1934, -106.8162, down to the Mill St. bridge.

W-04F: The parking here is at the upstream edge of a long

stretch of river that is one giant inside bend from 39.2007, -106.8263 to the Castle Creek confluence at 39.2050, -106.8329.

Locale: Aspen
Land Type: river running through a resort town
Land Manager: Town of Aspen

Key Regulations:
• Pans only.
• Fill any holes with your tailings.
• Only dig downward into the riverbed, never into a bank.
• Parks are open from dawn to dusk.

Nearby Attractions & Accommodations:
Well, it is Aspen, so maybe a fancy meal to celebrate your gold discoveries?

Site Number: W-05A-E
Site Name: Basalt Town Parks

The town of Basalt is essentially split into two pieces by the canyon walls. These are referred to as West Basalt and East Basalt but both are one city. This very picturesque setting makes for a great place to spend some time in the river.

Local Hints & Cautions:
• See points included in the prior site.

Gold Finding Tips:
• See points included in the prior site.

Getting There:
W-05A: Ponderosa Park: Parking on Basalt Ave. at 39.3656, -107.0340 (just north of the roundabout at 136 Emme Rd.)

W-05B: Confluence Park: At the confluence of the Frying Pan River the Roaring Fork, just across the river on the opposite side of Basalt Ave. from the prior site. Use the same parking area. See Roaring Fork River sites listed below for more prospecting opportunities!

W-05C: Midland Addition Riverfront Park: Park along Midland Ave. on the east side of the Basalt Library (39.3667, -107.0371).

W-05D: Old Pond Park: Use the street-side parking at 22830 Two River Rd. and walk into the park. (At the time of this publication, there is park construction going on between this site and the Midland Ave. bridge. There may be more/better parking provided there soon.)

W-05E: CDOT rec path access: unlike the other sites here, this one is entirely on CDOT property so electric high bankers are fine and gas-power too (if at least 50 feet from CDOT structures).

Boundaries: For all sites, stay on the park side of the river, the other side is generally private.

W-05A: Ponderosa Park & River Preserve: Upstream access starts at 39.3618, -107.0295 (stay downstream of the aeriation waterfalls) and extends downstream to 39.3659, -107.0338 at the Basalt Ave. bridge. It may also be possible to park at a couple spots along Emma Rd. (watch carefully for relevant signage) such as the river path access at 39.3627, -107.0303 (which is the upstream end of Emma Rd. and the upstream edge of the prospecting area) or about mid-way downstream at 39.3634, -107.0316.

W-05B: Confluence Park: From the Basalt Ave. bridge to 39.3668, -107.0347.

 W-05C: Midland Addition Riverfront Park: From the Midland Ave. bridge downstream to 39.3674, -107.0384.

W-05D: Old Pond Park from upstream at the Midland Ave. bridge (right across the river from Midland Addition park), to 39.3679, -107.0401 (do not continue downstream, the remaining park area is owned by the county and they do not welcome prospecting).

W-05E: CDOT rec path access: From upstream, east of the bridge at 39.3688, -107.0666 to downstream at 39.3703, -107.0714 (watching the longitude is enough here since the river runs east to west). Also, take the pedestrian bridge to the south side of the river to access that side from 39.3685, -107.0666 to the area under the highway bridge and 50 feet downstream of the bridge.

Locale: Basalt, between Glenwood Springs and Aspen
Land Type: river running through a valley
Land Manager: Town of Basalt, CDOT

Key Regulations:
- Pans and hand tools only.
- Fill any holes with your tailings.
- Only dig downward into the riverbed, never into a bank.
- Parks are open from dawn to dusk.

Nearby Attractions & Accommodations:
Camp east of town, up the Frying Pan River (see next sites) or go fishing in the river in town. Enjoy a hike in the hills nearby.

Site Number: W-06A, B
Site Name: Glenwood Springs city parks

The town of Glenwood Springs is in both this chapter and Chapter S for the prospecting access along the Colorado River.

Local Hints & Cautions:
- Stay on the park side of the river at each site, the opposite side of the river is private property in every case.

Gold Finding Tips:
- As mentioned in other sites along.

Getting There: To get to this general area, get off of I-70, Exit 116 for CO-82 East toward Glenwood Springs/Aspen and drive through the cool downtown Glenwood Springs area.

W-06A: Three Mile Creek Park: Three Mile Creek along the trail, access at 39.5107, -107.3190, access via Midland Ave. just south of 3 Mile Rd., on the east side; parking for 8 cars. Alternative parking on Mountain Drive, just around the corner to the SE from the first parking spot, this one is at 39.5097, -107.3178 with parallel parking along Mountain Drive.

W-06B: O'Leary & River Trail Dog Park: 1701 Riverside Dr.

Boundaries:
W-06A: Three Mile Creek Park in 3 Mile Creek from the parking lot down to the river. In the river from way upstream around the bend at 39.5125, -107.3103 to downstream past the creek to 39.5153, -107.3207 for a total of 0.6 miles of the river.

W-06B: From far south of the parking area, at 39.5299, -107.3269 to just north of the pedestrian bridge at 39.5387, -107.3304; however, I wouldn't bother going south of the big inside bend near the fenced dog park area.

Locale: Glenwood Springs
Land Type: river running through a semi-urban area
Land Manager: Glenwood Springs

Key Regulations:
• See prior site for details.

Nearby Attractions & Accommodations:
Visitors to Glenwood Springs have been enjoying the famous hot springs pool and the classic Grand Ave. downtown strip for well over 100 years. Enjoy a post-digging soak in the hot springs and a great meal on Grand Ave. to finish your day.

Frying Pan River Gold
The next several sites are along Frying Pan Road east of Basalt. Each of these sites are on the Frying Pan River itself. Although the bulk of the gold produced commercially in the Aspen/Basalt area came from the upper Roaring Fork River,

there is also gold here.

Site Number: W-07
Site Name: Chapman Campground

The sprawling Chapman USFS campground has the Frying Pan River running right through it. In addition, the claim withdrawal area here is quite large providing plenty of prospecting area despite some of it being boggy.

Local Hints & Cautions:
- See the comments just below on parking options along Frying Pan Road to minimize the hiking required at this large prospecting area.

Gold Finding Tips:
- During wet periods the upstream portion of this area is very boggy and would most likely be difficult to prospect.

Getting There: Take Frying Pan Rd. east out of Basalt. Turn right off of Frying Pan Rd. at Chapman Campground Rd. at 39.3207, -106.6445. Those wanting to avoid the day-use fee at the campground can try parking along Frying Pan Rd. at locations such as 39.3095, -106.6279 near the upstream end, 39.3166, -106.6360 or 39.3190, -106.6403 toward the downstream area.

Boundaries: The upstream boundary is 39.3041, -106.6194 which is not too far southeast of the end of the "Loop D" campground road. The downstream edge of the prospecting area is at 39.3215, -106.6507, with private property west of that point.

Locale: east of Basalt in the Frying Pan River valley
Land Type: river running forested valley
Land Manager: USFS

Key Regulations:
- No gas-powered equipment near the campground.

- Do not disturb erosion control rocks or structures along the bank of the river near the road.

Nearby Attractions & Accommodations:
Enjoy the reservoir just downstream or explore local trails. This is a beautifully remote area.

| Site Number: W-08 |
| Site Name: Deerhamer Campground |

The Deerhamer USFS campground is on the east end of Ruedi Reservoir. The claim withdrawal area here is quite large due to the existence of the reservoir. PLO 3500 reserved space for the reservoir and made a large area here unclaimable. The campground is on the shore of the reservoir at the point where the river enters the reservoir. Access here is ideal.

Local Hints & Cautions:
- See the comments just below on parking options along Frying Pan Road to minimize the hiking required at this large prospecting area.
- If there are staff in the U.S. Reclamation Bureau office near the campground, check in with them and get their approval on your prospecting plans. This will avoid misunderstandings later.
- This area is very popular, at times, with the fly-fishing community. Do not approach or speak loudly to someone while they are fly-fishing. Of course, if they approach you, be friendly and ask how the fishing is going!

Gold Finding Tips:
- As I have commented in describing other reservoirs, the upstream edge of a reservoir is a place where gold accumulates as it migrates downstream in the river. When the fast-flowing river hits the stiller waters of the reservoir, all the heavies drop out immediately. Since that meeting point changes as the reservoir levels change, there will a stripe of waterway that is sometimes under the reservoir and sometimes flowing as part of the river. This area can be

a true honey hole for fine flood gold. If you are lucky enough to be here when the reservoir is lower, sample where the river is flowing through the reservoir bed.

Getting There: Take Frying Pan Rd. east out of Basalt. Turn right off of Frying Pan Rd. at Lake View Rd. at 39.3612, -106.7368. Those wanting to avoid the actual campground, to avoid a day-use fee, can try parking along Frying Pan Rd. at locations such as 39.3616, -106.7327 or just west of there at 39.3612, -106.7361. Both spots have dirt routes toward the river.

Boundaries: The upstream boundary is 39.3628, -106.7298 which is well east of the campground, with Frying Pan Rd. paralleling the water as it flows toward the Ruedi Reservoir. The downstream boundary is wherever the river runs into the reservoir.

Locale: east of Basalt in the Frying Pan River valley
Land Type: river running semi-forested, wider valley
Land Manager: USFS, Bureau of Reclamation

Key Regulations:
- No gas-powered equipment near the campground.
- Do not introduce new material into the reservoir. (but take as much away as you like, making a reservoir bigger is helpful lol!)
- Do not dig near the edge of the reservoir where the sand and gravel bed of the reservoir meets the plants that form groundcover as this can lead to erosion when the reservoir is full.
- Do not disturb erosion control rocks or structures along the banks of the reservoir or river.

Nearby Attractions & Accommodations:
Enjoy the reservoir or explore local trails. This is a beautiful area. If you prefer a campsite with shade, head to one of the campgrounds upstream (east) of here instead. The general store in the little town of Meredith here has basic foodstuffs and some camping supplies.

- Rocky Fork Creek Day Use Area just below the Ruedi Reservoir Dam, turn off of Frying Pan Rd. at 39.3631, -106.8314.

Site Number: W-09
Site Name: Rocky Fork Day Use Area

It may seem like an odd name for this spot, since the day use area isn't on Rocky Fork Creek, but this area gives access to the Frying Pan River just downstream of Ruedi Reservoir.

Local Hints & Cautions:
- This area is very popular, at times, with the fly-fishing community. Do not approach or speak loudly to someone while they are fly-fishing. Of course, if they approach you, be friendly and ask how the fishing is going!

Gold Finding Tips:
- Don't worry about being just below the reservoir, the gold found its way here for many millennia before the dam was built.
- Do keep an eye on water levels and any signage about varying water levels since the outflow from the dam may change.

Getting There: Take Frying Pan Rd. east out of Basalt. Turn right off of Frying Pan Rd. at Rocky Fork Day Use Area Road at 39.3631, -106.8314. From there explore the access road on both sides of the river heading upstream but I recommend the road on the south side of the river because it leads to a great inside bend. Heading downstream on the south side of the river will also lead to some nice inside bends in the river. To access the downstream end, which is well past the end of the day use road, park on the shoulder of Frying Pan Rd. at 39.3614, -106.8405; be sure to cross the road carefully, there is a blind curve here.

Boundaries: The upstream boundary is the output of the dam. The downstream edge is at 39.3609, -106.8419 which is very, very close to the road.

Locale: east of Basalt in the Frying Pan River valley
Land Type: river running forested valley
Land Manager: USFS, Bureau of Reclamation

Key Regulations:
- No gas-powered equipment here due to the fishing.
- Do not disturb erosion control rocks or structures along the bank of the river near the road.

Nearby Attractions & Accommodations:
Enjoy the reservoir just upstream or drive all the way east on the road that starts out on the south side of the river. This road is called Rocky Fork Creek Road, and it leads up past the dam to follow the creek uphill a bit to a picnic spot with a wonderful view.

Oh, let's not forget to show our hard rock miners some love too!

End Notes

Notes to K-12 Educators in public schools:

If you would like to teach your students about gold prospecting in Colorado, feel free to copy anything from this book for use in the classroom.

If you would like to teach your kids to gold pan, local prospecting clubs are often happy to provide volunteer instructors. A full list of active clubs across Colorado is available on my website at www.findinggoldincolorado.com so you can easily track down the nearest club.

Low-cost supplemental curricular and classroom materials including videos, posters, gold pans, instructional booklets, and gold paydirt can be obtained from the Minerals Education Coalition via their website at www.MineralsEducationCoalition.org and via their online store.

Teachers seeking to organize gold rush related fieldtrips can reach out to the local gold prospecting community to find volunteers via our Finding Gold in Colorado Facebook group. I'd recommend 4th grade as the youngest age to consider this activity. Many schools choose to include a field trip related to mining history and gold panning as part of the 4th Grade Colorado curriculum.

Acronym and Abbreviation Guide:

4WD: Four-wheel drive (vehicles)
BLM: Bureau of Land Management
CC&V: Cripple Creek & Victor
CDOT: The Colorado Department of Transportation
CG: Campground
CGC: Colorado Gold Camp club
CPW: Colorado Parks and Wildlife

DCOS: Douglas County Open Space
EPA: Environmental Protection Agency
GPOC: Gold Prospectors of Colorado, club in Colorado Springs
GPAA: Gold Prospectors Association of America, has several chapters in Colorado
GPOR: Gold Prospectors of the Rockies, club in metro-Denver
GPS: Global Positioning System, satellite-based tool which identifies latitude and longitude coordinates
PLO: Public Land Order, a federal land management regulation authorized by congress
RV: Recreational Vehicle
USFS: United States Forest Service
USGS: United States Geological Survey
VRBO: Vacation Rentals by Owner
WFMU: Western Federation of Miners Union

Road Naming standard in this book:

US-xx: United States highway designations
CO-xx: Colorado State highway designations
CR-xx: County Roads
FR-xx: US Forest Service Roads

Vocabulary:

Braided: A riverbed condition in which the river splits into several channels of water flow with low lying islands in between. This typically occurs in a situation where the land flattens out and the riverbed widens. This combination allows the water flow to slow down and thus to drop its burden of sediment. Often the upstream portion of a braided area will be richer in gold as a result of this slowing and dropping of sediments. Note: during flood conditions, the braided nature of the riverbed will typically not be visible since the entire riverbed will be submerged temporarily.

Claim: also, Federal Mining Claim: A federal mining claim is a selected parcel of Federal land, valuable for a specific mineral deposit or deposits, for which you have asserted a right of possession under the General Mining Law. Your right is restricted to the development and extraction of a mineral deposit (per the BLM)

Color: A piece of gold regardless of size but often referring to very small specks.

Coyote Hole: A hole burrowed into a bank following a layer of paydirt. A very dangerous way to dig a hole. Generally, not permitted in public access prospecting areas as it is a hazard to both the prospector and to those who come along later.

Inside Bend: When a waterway snakes across the landscape, it creates inside bends. These are the areas where the waterway arcs around a peninsula of land. Often the peninsula is formed by the action of the water itself. This action, in which lower moving water deposits it burden of sediments as it slows down, creates the peninsula shape. That slowing also means the flowing water will drop any gold it is carrying in the flow up until that moment. As a result, gold tends to accumulate on the inside of a river bend.

Mesh: Number of holes per linear inch in a sand and gravel classifying screens. Larger mesh numbers mean the material that can flow through the classifier is smaller. For example, material that flows through a 20-mesh classifier but not a 30-mesh classifier would be larger than material that did get through the 30-mesh classifier. Also, anything caught in a 20-mesh classifier would be referred to as "+20 material" and anything that falls through would be "-20 material."

Mining: A large or small-scale commercial operation to recover valuable minerals. Mining is only allowed on a federal mining claim or on private land with the permission of the owners. Formal permits are typically required from multiple government agencies. Mining is not allowed at any of the sites described in this book.

Placer: a deposit of gold or other valuable mineral created by a waterway, or by beach wave action. The mineral will be mixed with other material which has been eroded into the waterway. Over time, the water will tend to sort all of the material by density, causing the gold to concentrate in certain areas often called pay streaks. As waterways move across the landscape over time, a placer can end up quite a distance from the water. These dry placers are often referred to as bench placers.

Prospecting: Searching for valuable minerals. This can be done in areas open to mining claim as well as in other areas (where not banned by law or lawful regulation). Prospecting can be done in areas "Withdrawn from Mineral Entry" and in Federal Wilderness Areas but with certain restrictions on equipment used unless an explicit law is written to ban Prospecting.

Recreational Gold Mining: A vague term first used in the 1960s, referring to small scale/casual placer mining or casual placer prospecting or even just panning. It seems to me and many others to be absurd to connect the words recreational and mining since an activity may be one or the other but not both at once. Whatever it is supposed to be called, the "recreational", or casual and small-scale mining boom started in 1979 when changes in the gold ownership and valuation laws in the U.S. led to a gold price spike from $35 peaking at about $800/ounce.

Recreational Panning and Prospecting: Panning and other limited forms of prospecting are welcomed in many public areas owned and managed by local governments and on restricted-use lands managed by federal agencies. In these places the activity is considered 'recreational' since no commercial scale activity is allowed and there is no possibility of the prospecting activity leading to a claim being located (aka filed). There is no such thing as recreational prospecting on federally managed lands open to claim location. Think for a minute about what you would do in such a situation if you found a rich placer deposit or a vein, you would file a claim on it and get rich right?! This means that unlike hunters and anglers who accept limits on their activities (2 fish per day, one bull elk, whatever,) when on relevant lands, we have unlimited possibilities to expand our casual, weekend prospecting into something life changing. Nothing 'recreational' about that right?

Tailings: the sand, gravel, and rocks leftover after processing material to capture the gold in it.

THANKS FOR READING MY BOOK! - KEVIN

BIBLIOGRAPHY

<u>Books & Scholarly Works</u>:
Gold Occurrences of Colorado by Mark Davis and Randall Streufert, Colorado Geological Society, 2011
Gold Panning and Placering in Colorado: How and Where, by Ben H. Parker, Jr., 1961, 1974, 2009, Colorado Geological Survey, Department of Natural Resources
Gold Placers of Colorado, by Richard Lampright, Iron Fire Publications, 2010

<u>New resources used for this book</u>:
Boomtown to Ghost Town: The Story of Fulford, by Richard Perske, 2015
Holy Cross City: the gold camp built on hope, by Kathy Heicher, Vail Valley Magazine, 2023
Mountain Biking Colorado's Historic Mining Districts, by Laura Rossetter, 1991
Unique Ghost Towns and Mountain Spots, by Carolyn Bancroft, 1961
Where is the Gold on the Colorado River, by Dell R. Foutz PhD

<u>Marketing Materials and Tourist Information</u>:
Multiple publications, by local history associations and chambers of commerce
Multiple publications, by local Parks & Recreation or Open Space districts
Multiple publications, by the local ranger districts

<u>Websites</u>:
Multiple Colorado County Assessor websites for information on land ownership.
Multiple government websites with information on campgrounds and local recreational opportunities.
www.mylandmatters.org for land ownership and master title plat research materials.
Flaglane.com for the image of the Utah flag
Wikipedia, for background information on places and people.

INDEX (place, chapter)

Pueblo, H

Mancos, O

Maybell, R

McCoy, S

Monarch Spur RV Park, H

Radium, S

Salina, B

Sedgwick, A

Silt, Q

Silverton, O

Steamboat Springs, R

Sterling, A

Telluride, M

Thornton, A

Trinidad, V

Twin Lakes, G

Westwater Utah, Q

Whitewater, Q

Windsor, U

Wolcott, T

ABOUT THE AUTHOR

Kevin Singel caught gold fever at the age of six when his father read a National Geographic magazine article which included gold prospecting at Vic's Gold Panning in Blackhawk, Colorado to him. He promptly went out and dug a 'gold mine' in the front yard (good thing there wasn't a lawn yet!)

When he arrived in Colorado in 1989, one of his first weekends was spent learning to pan and buying his own gold pans.

In recent years, his interest has become a passion with this series of books as a tangible result. Kevin also authors the website www.findinggoldincolorado.com and manages the Facebook group 'Finding Gold in Colorado' as well. He encourages all those interested in casual Colorado gold prospecting to join the Facebook group to get advice and current events information. Joining the Facebook group and subscribing to the website are also the best ways to get updates on this book, to buy the original *Finding Gold in Colorado: Prospector's Edition* book and to keep up with Kevin's future work.

Kevin looks forward to doing more to support the small Colorado prospector and hopes the gold prospecting community enjoys this book.